❖

DAMS and
DEVELOPMENT

Dams and Development

Transnational Struggles for Water and Power

SANJEEV KHAGRAM

CORNELL UNIVERSITY PRESS

ITHACA AND LONDON

First published 2004 by Cornell University Press
Cornell Paperbacks, 2004
Printed in the United States of America

Library of Congress Cataloging-in-Publication Data

Khagram, Sanjeev.
    Dams and development : transnational struggles for water and power / Sanjeev Khagram.
        p. cm.
    Includes bibliographical references and index.
    ISBN 0-8014-4228-1 (cloth : alk. paper) — ISBN 0-8014-8907-5 (pbk. : alk. paper)
    1. Human ecology—India—Narmada River Region.   2. Environmental degradation—India—Narmada River Region.   3. Dams—India—Narmada River Valley.   4. Water resources development—Environmental aspects—India—Narmada River Valley.   5. Dams—Environmental aspects.   6. Dams—Economic aspects.   7. Dams—Social aspects.   8. Human ecology—Cross-cultural studies.   I. Title.
    GF662.N37K46 2004
    333.91′00954—dc22

                                                                                        2004007162

Cornell University Press strives to use environmentally responsible suppliers and materials to the fullest extent possible in the publishing of its books. Such materials include vegetable-based, low-VOC inks and acid-free papers that are recycled, totally chlorine-free, or partly composed of nonwood fibers. For further information, visit our website at www.cornellpress.cornell.edu.

Cloth printing   10 9 8 7 6 5 4 3 2 1
Paperback printing   10 9 8 7 6 5 4 3 2 1

❖

To my son, Ram, and all the children of the world.
Never lose faith that you can achieve
the seemingly impossible.

CONTENTS

❖

FIGURES

TABLES

❖

ACKNOWLEDGMENTS

This book was touched by people spread across five continents and is the culmination of a journey that has taken nearly a century.

In many ways, work for this study began when my ancestors emigrated from Gujarat to East Africa in the early 1900s. My own quest for knowledge was inspired by one of these family members, my grandfather Damji-bhai Khagram, whose passion for learning was reflected by his ability to speak six languages, including Sanskrit. This book is a small attempt to honor him, may his soul rest in peace, and our family name.

My family's expulsion from Uganda in 1972 and subsequent relocation to the United States was a critical juncture in this journey. This book is a testament to more than thirty long years of struggle by my parents Ramesh and Urmila. They sacrificed so that my sister and I could have an "education that would contribute to the world." This book is one symbol of their success.

To my older sister, Shilpa, I offer my deep gratitude. She was always the first to take a path that had never been walked by any of us before. Her courage to lead the way made each step much easier for me than it would otherwise have been.

To Regina Segura-Khagram I offer my love and respect. More than any other person, you shared the roller coaster of this book with me. It certainly could never have been written without your patience and perseverance.

A book such as this could also not be written without the unwavering support of countless mentors, teachers, colleagues, and friends. I am deeply grateful to Terry Karl, Philippe Schmitter, Larry Diamond, John Meyer, Akhil Gupta, Lynn Eden, David Holloway, Stephen Krasner, Kathryn Sikkink, Martha Finnemore, Jim Riker, Len Ortolano, Pan Yotopolous, Sonia Alvarez, Allison Brysk, Suzanne Shanahan, Chuck Call, Tracy Fitzimmons, Marvin Peguese, Federico Besserer, Alnoor Ebrahim, David Valdez, Isaac Solotaroff, Jendayi Frazer, Roya Razaghian, Flavia Braga, Carlos Vainer, S. Parasuraman, Taeku Lee, Archon Fung, Afreen Alam, Adanna Scott, Sarah Alvord, Peggy Levitt, Shawn Bohen, Mark Moore, Frances Kunreutler, Srilatha Batliawala, David Brown, several anonymous reviewers and so many others for their invaluable time, advice, and encouragement. Research and writing for the book could not have been possible without the generous support of the MacArthur Foundation, Stanford Center for International Security and Cooperation, American Institute of Indian Studies, Institute of World Politics, Hauser Center for Non-Profit Organizations, Carr Center for Human Rights Policy, Harvard Center for International Development, Kennedy School of Government, David Rockefeller Center for Latin American Studies, and the Center for International Affairs.

In this book, I have attempted to offer a glimpse into the tremendous energy, imaginativeness, and courage of the millions who struggle to make our world more democratic, socially just, culturally nurturing, and environmentally sustainable. To all those in India, Brazil, South Africa, Indonesia, China, the United States, and more than thirty countries across the world who have offered me their insights, experiences and wisdom, I hope that this book at least minimally reflects what you have shared with me. For all the errors that may remain, I not only ask for your understanding but take full responsibility. I pray also that this study provides a vision of how the seemingly impossible can become reality, a lesson about transformation I have learned from all of you and will never forget.

SANJEEV KHAGRAM

Harvard University
February 2004

❖

DAMS and
DEVELOPMENT

❖

Transnational Struggles for Water and Power

We need large dams and we are not going to apologize for it. Those in the developed countries, who already have everything put stumbling blocks in our way from the comfort of their electrically lit and air conditioned homes. . . . The Third World is not ready to give up the construction of large dams, as much for water supply and flood control as for power. . . . Hydropower is the cheapest and cleanest source of energy, but environmentalists don't appreciate that. Certainly large dam projects create local resettlement problems, but this should be a matter of local, not international concern.

> FORMER PRESIDENT THEO VAN ROBBROEK of the International
> Commission on Large Dams (ICOLD).[1]

To persuade Third World governments to abandon plans to build water development schemes, to which they are often totally committed, is very difficult. Nevertheless, every effort must be made by local environmental groups to do so. If necessary, they should resort to nonviolent direct action at the dam site. We in the West can best prevent the construction of further dams by systematically lobbying donor governments, development banks, and international agencies, without whose financial help such schemes could not be built. Indeed, we call on those organizations herewith to cut off funds from all large-scale water development schemes.

> EDWARD GOLDSMITH AND NICHOLAS HILDYARD, editors of *The Ecologist.*[2]

The Narmada projects are irrigation, power, modernization. The projects are development.

> Y. K. ALAGH, former Chairman, Narmada Planning Group,
> Government of Gujarat, India.[3]

Development? What sort of development is this? Development with destruction?

> MEDHA PATKAR, leader, Narmada Bachao Andolan—
> Save the Narmada Movement.[4]

Projected to generate thousands of megawatts of power, irrigate millions of hectares of land, and supply drinking water to hundreds of villages, India's Narmada Projects represent a promise for plenty to its proponents. If completed, these projects would undoubtedly constitute the largest river basin scheme in India.[5] The Narmada Projects are also expected to submerge thousands of villages, displace millions of mostly peasant and "tribal" people, and destroy tens of thousands of hectares of forest lands.[6] Opponents thus charge that they would be one of the greatest planned social and environmental tragedies in the world.[7]

Not surprisingly, the Narmada Projects have had a long and conflict-ridden history. After more than three decades of investigations and planning, Indian authorities finally sanctioned the development initiative in 1978.[8] The gargantuan scale of the river basin scheme ultimately approved—with 3,000 "small dams" and 30 "big dams"—including the 455-foot-tall Sardar Sarovar and 865-foot-tall Narmada Sagar "major dams"—drew the immediate interest of the World Bank.[9] The Bank commenced its formal support within the year, at first for the Sardar Sarovar major dam component of the broader Narmada Projects. This quickly attracted further backing from the United Nations Development Program, Japan, and other foreign donors. At the time, proponents confidently asserted completion of the Sardar Sarovar Project in less than a decade.

But fifteen years later, in a dramatic turn of events, construction on the Sardar Sarovar Project was stalled, and the entire set of projects planned for the Narmada River Valley seemed imperiled. Japan and other donors had been compelled to withdraw their support due to a transnational campaign waged against the projects by a grassroots social movement with the support of nongovernmental organizations and allied groups from across India and all over the world.[10] Under severe pressure from these opponents, the World Bank acquiesced to the first ever independent review of a project it was funding.[11] The review team produced a highly critical report, and a crisis within the Bank ensued.[12] Besieged by anti-dam proponents at home and losing credibility internationally, domestic federal authorities grudgingly announced that the Government of India would forego hundreds of millions of dollars of World Bank funding. Then, in a blow to domestic proponents, India's Supreme Court ruled to stall construction on the Sardar Sarovar Project indefinitely.

The struggle over India's Narmada Projects is not an isolated one. Rather, it is part of a historical trend of mounting domestic and transnational contestation over big dam projects that has spread to all regions of the world. Over the past quarter century, opponents of big dams have contributed to the reform, postponement, cancellation, and even decommissioning of these projects in industrialized countries, such as the United States, Sweden, and France; in the former Communist bloc, Soviet successor states and Eastern Europe; and across the third world from Chile to Namibia to Nepal. Conflicts over the construction of big dams have grown into intense policy debates in numerous countries around the world.

At the international level, big dams have been at the center of highly visible campaigns to substantially reform the procedures and practices of multilateral organizations (particularly the World Bank), bilateral aid and export credit agencies, and multinational corporations. Most recently, the coordinated activities of an emerging transnational anti-dam advocacy network contributed to the creation of a World Commission on Dams. An innovative independent, multi-stakeholder experiment in global governance, the Commission's mandate entailed the first ever comprehensive review of the development effectiveness of big dams around the world and the formulation of new global norms for the planning, implementation, operation, and decommissioning of these projects.

Why have historically weak and marginalized actors been increasingly able to prevent far more powerful interests and organizations from constructing big dams? How has the governance of these projects been altered from the local to the international level even when they have been built? What are the broader empirical implications of these changing dynamics around big dams for democracy and development? Do the answers to these queries contribute to theoretical advances in the social sciences?

In this book, I argue that the unpredicted emergence and unexpected strength of transnationally coordinated action—constituted primarily by nongovernmental organizations and social movements—has dramatically altered the dynamics surrounding big dams around the world from the local to the international levels. Critics and opponents of these projects have been strengthened by globally spreading norms and principles regarding human rights, indigenous peoples, and especially the environment, among others.[13] The gradual institutionalization of these norms and principles into the procedures and structures of states, international agencies, private sector companies, and other prominent organizations has partially been the result of the activities, and substantially contributed to the effectiveness, of these actors.

But transnationally allied opposition to big dams does not have the same impact everywhere. First, and perhaps most importantly, outcomes are most likely to be altered when domestic communities and social movements capable of generating sustained mass mobilization and linked to transnational advocacy efforts contest these projects. Second, big dam critics and the range of tactics they employ are likely to be much more effective in democratic institutional contexts. These regimes offer greater opportunities to organize and gain access to decision-making processes, and significantly reduce the ability of big dam proponents to violently repress contestation.

These changing transnational dynamics surrounding big dams further highlight a broader transformation that has taken place in the political economy of development. Behind the intensified and transnationalized struggles over big dams lie deep struggles between competing visions and models of development. A range of powerful, transnationally allied groups and organizations have historically promoted the construction of these projects: politicians, bureaucrats, landed classes and industrialists, multinational corporations, the World Bank and other international organizations, as well as transnational professional associations of engineers and scientists. The continuing interaction among these big dam proponents generated an informal international "big dam" regime by the 1950s, legitimating and naturalizing the construction of these projects around the world. A central underlying aspect of this regime were the deeply rooted norms and principles, which taken together promoted a vision of development that remained hegemonic for the subsequent half-century. This vision equated development as a large-scale, top-down, and technocratic pursuit of economic growth through the intensive exploitation of natural resources.

Hence, critics of big dam building have had to challenge not only powerful interests and dominant institutions, but also hegemonic ideas about development in their struggles against big dam projects.[14] Indeed as I demonstrate in this book, the vision that legitimized and naturalized particular processes and outcomes of development is being contested by a novel set of transnationally allied actors: previously marginalized but increasingly mobilized grassroots groups and social movements along with rapidly proliferating nongovernmental organizations promoting social justice, human rights, cultural diversity, and environmental preservation, among others. As a result, an alternative vision of development that advances bottom-up and participatory processes directed toward socially just and ecologically sustainable outcomes has been gaining greater support and is becoming increasingly powerful around the world.

Nowhere have the conflicts between these competing visions of development been more vividly displayed than in the intensified transnational contestation over big dam projects. Big dams have been characterized contrastingly as examples of modernization and destruction, temples and tombs, as progress and injustice. On the one hand, the massive scale of these projects, and their seeming ability to bring powerful and capricious natural forces under human control, has given them a unique hold on the social imagination. Perhaps more than any other development initiative, big dams have symbolized the progress of humanity from a life controlled by nature and tradition to one in which nature is ruled by technology, and tradition supplanted by science. On the other hand, big dams have more recently become symbols of the injustice of humanity through the untold destruction of nature, and the sacrifice of diverse cultures to inappropriate science and technology in the name of progress.

The purpose of this book is not to determine whether one of these development visions, big dam building generally or a specific project, is desirable or not. Rather my overarching goals are to describe and explain the changing dynamics surrounding big dams, and the broader empirical and theoretical implications of these changes. Partly in order to successfully do this and partly as a result of doing this, it is necessary to recognize that big dams have been socially constructed during the 20th century as premier development activities and symbols: as a result, the changing dynamics surrounding big dams offer insights into changing dynamics of development more generally. But it must also be acknowledged that the changing dynamics surrounding big dams both shape and are shaped by changing dynamics of development more broadly. Analyzing, understanding, and theorizing this complex set of relationships between big dams and development is the task to which I now turn.

THE RISE AND DECLINE OF BIG DAMS GLOBALLY

The trajectory of big dam building over the last 100 years provides a striking illustration of a broader transformation in the transnational political economy of development. The rise and global spread of big dam building is clearly a 20th-century phenomenon. In 1900, there were approximately 600 big dams in existence, many of the oldest of which were built in Asia and Africa. The figure grew to nearly 5,000 big dams by 1950, of which 10 were major dams. By the year 2000, approximately 45,000 big dams, including approximately 300 major dams, had been constructed around the world (see figure 1.1)! Thus, over 90 percent of big dams were built over the last forty years.[15]

Big dam building also spread or was transferred from a small number of river basins to all regions of the world during this period: these projects have now been erected in at least 140 different countries.[16] Great Britain had more than half of the world's big dams at the turn of the last century.[17] In 1902, British authorities constructed the low Aswan Dam on the Nile River, subsequently raising it twice to over 100 feet in height by 1933.[18] The Desprostoi on the Dnieper River in the Soviet Union was built a year earlier; it was the world's first major and most powerful hydropower dam at the time.[19] By the 1930s, the United States Bureau of Reclamation (BuRec) had built over 50 big dam projects and had commenced work on the mammoth Hoover Dam.[20] Completion of the Hoover Dam, and the establishment of the Tennessee Valley Authority (TVA), which built 38 large dams before 1945, followed by the construction of the even

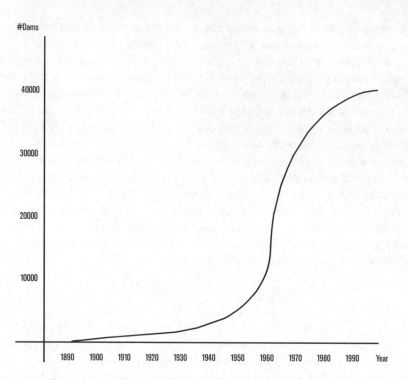

Figure 1.1. Cumulative Number of Big Dams Constructed Worldwide

larger Shasta and Grand Coulee projects, heralded the big dam era in the United States.[21] Similar efforts were under way in other countries around the world, particularly in the former Soviet Union and in Western Europe.[22]

Transnationally allied proponents of big dam projects increasingly became more linked during these early years, contributing to the formation of an informal international big dam regime. In 1929, an array of engineers, builders, and bureaucrats established a transnational professional association called the International Commission on Large Dams (ICOLD) to collect information and coordinate the exchange of knowledge about big dam building around the world.[23] Beginning in the 1930s and 1940s, big dam-building bureaucracies, including national agencies such as BuRec or the former Soviet Union's Hydrological Planning Agency and river valley organizations like the TVA, proliferated around the world. The wave of decolonization and state-formation that began in South Asia in 1947 substantially contributed to the spread of these institutions across the third world. By 1955, BuRec Commissioner Michael Strauss claimed that "the American concept of comprehensive river basin development," involving the creation of these bureaucratic agencies and the building of big dams, had "seized the world imagination."[24]

The World Bank, which soon emerged as the premier multilateral development agency, took the lead in supporting the construction of big dams across the third world. The first loan it ever awarded outside Europe underwrote three big dams in Chile, and its first loans to fourteen other developing countries were for similar projects. Promotion of comprehensive river basin planning was also taken up in earnest.[25] As noble-prize winning development economist Albert Hirschman later wrote, "any river valley scheme, whether it concerned the Sao Francisco in Brazil, the Papaloapan River in Mexico, the Cauca in Colombia, the Dez in Iran, or the Damodar in eastern India, was presented to a reassured public as a true copy of the Tennessee Valley Authority."[26] Since then, by one estimate, the Bank lent over $58 billion for more than 604 big dam projects in 93 countries. Of the 527 loans made by the Bank for big dams, more than 100 were the largest made to the borrowing country at the time of approval.[27]

The growth of a private dam industry and other development aid and credit agencies greatly augmented the power of big dam proponents, further contributing to the spread, growth, and legitimacy of big dam building across the third world. Prominent transnational corporations, such as Asea Brown Boveri, Siemens, GEC Alsthom, Kvaerner, Bechtel, Acres, and others, expanded rapidly during the 1950s and 1960s. Other multilateral organizations besides the World Bank, such as the specialized wings of the United Nations, especially the Food and Agriculture Organization and UN Development Program, and the Inter-American and Asian development banks, also increasingly played a major role in promoting big dams in developing countries. Bilateral aid and credit agencies like the British Overseas Development Administration (ODA) or the United States Export Import Bank increasingly became important funders of these projects, often in partnership with bureaucratic agencies in the third world, transnational corporations, and multilateral organizations.[28]

Despite the estimated more than US $2 trillion spent on big dam and major river valley projects over the last century, approximately one billion people currently do not have adequate supplies of water. By the year 2025, it is estimated that three billion people spanning fifty-two countries will be plagued by water stress or acute shortages of this precious life-sustaining resource.[29] Frictions between states over water resources are predicted by some to intensify, because nearly 40 percent of the world's river basins is shared by two or more countries.[30] Within and across countries, these conflicts are just as volatile, if not more so. As a result, water is now at the top of the global agenda.[31]

While conflicts over water have reemerged as a central concern, those over power are just as intense. The share in world energy consumption held by developing countries is expected to continue to rise from the current 25 percent to over 40 percent by 2010.[32] In anticipation of this growth, third world governments have focused increased attention on the energy sector. Despite the heavy investments on power projects and increasing private sector involvement in the sector, however, there are still more than two billion people in the world without stable sources of electricity.[33] Such shortfalls, exacerbated by the decline in hydropower big dam building and increasing concerns about the contribution of fossil fuel energy generation to global warming, present increasingly potent transnational political challenges.

Given the powerful set of big dam interests and institutions, along with the tremendous need for water and power around the world, the rapid decline globally over the last quarter century in the rate of big dam building is especially puzzling (see figure 1.2). The number of big dam projects completed per year had grown from approximately thirty in 1900 to nearly two hundred and fifty by mid-century. Thereafter, the rate exploded and peaked when, it was estimated, more than one thousand big dams were being finished annually by the mid-1960s. But even more dramatically, the number of these projects completed per year then fell precipitously to under two hundred by the turn of this century, representing a 75 percent drop in the construction rate of big dams in less than two decades (see figure 1.2)! The number of major dams completed over time similarly declined: during the 1970s, ninety-three of these mega-projects were constructed, while approximately twenty-five were built in the 1990s.[34]

Four main types of arguments can be offered to explain this puzzling trend: technical, financial, economic, and political. The technical argument highlights the decreasing availability of sites for big dam building to account for the falling completion rate around the world. Yet, 95 percent of big dams are concentrated in twenty-five countries in which more than one hundred have been built, while less than 2 percent are spread over the more than one hundred fifty other countries of the world where sites are still readily available, if not plentiful.[35] Only slightly more than 10 percent of the world's technically available hydropower potential has been developed by energy industry estimates.[36] During the 1990s, the number of big dams under construction remained flat at between eleven and twelve hundred per year, as did big dam-starts averaging three hundred per year.[37] Thus, there has been a diverging trend over time between the number of sites still available, as well as the number of dams being started and under construction, compared to the number of big dams actually being completed every year.

Financial and economic factors, such as shortages in available funding and the increasing relative viability of "conventional" alternatives for the services provided by big dams, are two other possible explanations for the decline in big

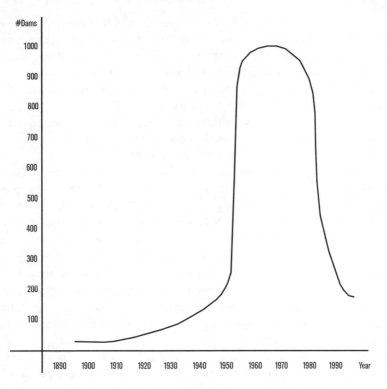

Figure 1.2. Average Number of Big Dams Completed Worldwide

dam building.[38] The 1973 and 1979 oil crises, worldwide recession during the 1980s, growth in indebtedness by many third world states, donor fatigue among foreign lenders, and a strategy shift toward privatization all contributed to the decreasing availability of public and international financing for these projects.[39] The lower costs of other conventional forms of energy production from time to time, such as natural gas power plants, relative to the development of hydropower, reduced the comparative economic feasibility of big dams.[40] The increasing time and cost overruns associated with big dam projects further detracted from expected financial and economic returns and improved the marginal utility of investing in other projects.[41]

These technical, financial, and economic factors have clearly made big dams less attractive, but they do not tell the whole story. Political-economic dynamics have increasingly contributed to the decreasing financial viability of and changing economic benefit-cost calculations regarding big dam building. Mounting public protests against big dams have caused time overruns and cost overruns.[42] Costs have also increased because big dam builders and authorities have been compelled to investigate and mitigate negative environmental and social effects, such as adequately compensating displaced peoples. When these environmental and social costs were internalized, benefit-cost or internal rate of return criteria for approving these projects were less likely to be satisfied. These requirements for the formulation, sanctioning, and implementation of big dams either did not exist or were not followed in the past.[43] Critics have additionally argued that many big dam projects are not financially or economically feasible and that less costly alternatives are available.

The technical argument of site depletion, moreover, has actually been as much a cause for the generation of political opposition to big dams as it has been a direct factor in the decline of big dam building. Beginning in North America and Western Europe, the growing loss of free flowing rivers associated with the damming of more and more sites sparked much of the initial organization and mobilization of critical domestic conservation groups (for more details see chapter 6). Indeed, the success of early anti-dam campaigns contributed to the growth of national environmental movements in numerous countries in these regions during the 1950s and 1960s.[44] The declining opportunities for big dam building in the first world, increasing demand and support for them across the third world as well as among development aid and export credit agencies subsequently drove proponents such as transnational corporations to shift more of their activities to developing countries. As a result, approximately two-thirds of the big dams built in the 1980s and three-quarters under construction during the 1990s were in the third world.[45]

But since the 1970s, coalescing from a multitude of struggles and campaigns waged from the local to the international levels, transnationally allied critics and opponents of big dams have dramatically altered the dynamics of big dam building across the world. Environmental nongovernmental organizations from the industrialized countries, in addition to those promoting human rights, the protection of indigenous peoples and other issues, have increasingly focused their energies on halting the global spread of big dam building. At the same time, directly affected people, grassroots groups, social movements, and domestic nongovernmental organizations in the rest of the world have increasingly empowered themselves and become more organized to reform or block the completion of big dams in their own river basins and countries, often by forging direct linkages with these like-minded foreign supporters. As a result, in contrast to the past when domestic technical, financial-economic, and political factors contributed relatively equally and interacted to cause the decline in big dam building, the primary explanation for these changing dynamics over time is increasingly political-economic, having to do with the shifting transnational power relations and meanings associated with the construction of big dams.

THE CHANGING TRANSNATIONAL POLITICAL ECONOMY OF DEVELOPMENT

A novel form of transnational dynamics constituted primarily by grassroots groups, social movements, and nongovernmental organizations from all over the world—and covering a wide range of issues, including security, trade, democratization, human rights, indigenous peoples, gender justice, and the environment—has become a common feature of world politics.[46] This study argues that a similar set of transnationally allied actors have also altered the political economy of development. These actors have been empowered by, and have also contributed to, the global spread and international institutionalization of norms in areas such as the environment, indigenous peoples, and human rights. But, as we shall see in the examination of the changing dynamics of big dam building across the third world, this combination of novel and increasingly influential transnationally linked action and globalizing norms has been most effective when undergirded by mass mobilization and multilevel advocacy in more democratic institutional contexts.

The tremendous proliferation of nongovernmental organizations around the world during the 20th century has been central to the new transnational political economy of development. Domestic nongovernmental organizations draw membership from, or are located in, only one country but can have either a domestic and/or an international focus, such as the United States-based Sierra Club or Cultural Survival.[47] In contrast, *transnational nongovernmental organizations,* such as Amnesty International or Greanpeace, draw membership from, and are active in, multiple countries and do not necessarily have allegiance to any particular state or society.[48] According to one estimate, the number of these transnational actors has exploded over the past century from 176 in 1909, to 832 in 1951, to 4518 in 1988. The ratio between transnational nongovernmental and international organizations, like the World Bank or the International Energy Agency, similarly rose during the period: from 5 to 1 in 1909, to about 8 to 1 by the 1950s, to more than 14 to 1 in 1988.[49]

The rapid growth in numbers of transnational nongovernmental organizations is especially visible with those groups publicly and nonviolently promoting social change in areas such as human rights, environment, and development (see table 1.1).[50] For example, in 1953 there were 33 transnational nongovernmental organizations working on human rights. The number at least doubled each decade between 1973 and 1993, growing from 41 to 190. Similarly, in the

Table 1.1. The Growth of Transnational Nongovernmental Advocacy Organizations

Issue area	1953	1963	1973	1983	1993
Human rights	33	38	41	79	190
Environment	2	5	10	26	123
Development	3	3	7	13	47

areas of environment and development, the number of transnational nongovernmental organizations grew gradually between 1953 and 1973, but then dramatically increased after that. Over the last decade, similar groups promoting development grew by over 500 percent, while the number of transnational environmental organizations skyrocketed (see figure 1.3)![51]

Not only have the numbers of transnational nongovernmental organizations grown, but transnational coalitions and networks linking nongovernmental organizations along with other allied actors, both domestic and transnational, have also increasingly been formed to promote social change in various issue areas.[52] *Transnational coalitions* are sets of actors linked across borders who publicly and nonviolently coordinate to formulate and implement a specific campaign, a shared strategy, or set of tactics.[53] *Transnational networks,* in contrast, are sets of actors linked across borders by some core shared normative concerns and dense exchanges of information and services.[54]

Transnational coalitions linking nongovernmental organizations, grassroots groups, and social movements across developed and developing countries are often formed for strategic purposes, in particular, to increase their potential to effect outcomes. In the third world, domestic actors are likely to construct transnational links with foreign or transnational counterparts to further their own goals, often when they face barriers in directly gaining access to, and leveraging, states, multilateral organizations, and transnational corporations. First world nongovernmental organizations engaged in issues that are international or transnational in nature often seek to build connections with third world allies to exchange information, increase their power, and gain legitimacy.[55] Although geographical distance, language and other cultural differences, varying organizational structures, and imbalances in resources pose substantial barriers to the effective functioning of these transnational coalitions, these types of interactions have nevertheless become more common.

Transnational coalitions, in contrast to transnational networks, can be formed when the strategic interests of different actors converge even if they are not bound together by deeply shared normative concerns. Thus, actors pro-

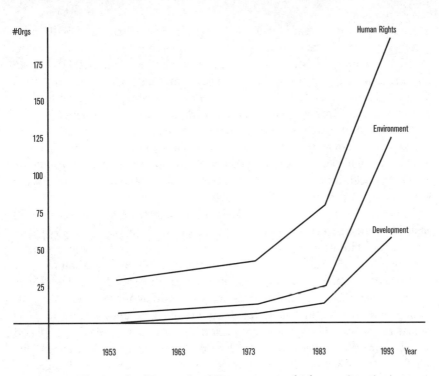

Figure 1.3. The Growth of Transnational Nongovernmental Advocacy Organizations

moting women's rights and those working on environmental conservation may coordinate a campaign to increase funding for women's education because it contributes to gender justice and decreases population growth. A campaign, or more likely a series of campaigns, waged by the same or different transnational coalitions can lead to the formation and continuation of a transnational network by contributing to the development of shared norms and dense exchanges of information and services. Over time, the existence of transnational networks makes the formation of additional links to other actors much easier. The existence of transnational networks also facilitates the construction of transnational coalitions in order to wage specific campaigns. Transnational coalitions and networks are often spawned, sustained, or invoked by nongovernmental organizations.

Thus transnational nongovernmental organizations, coalitions, and networks interact over time and increasingly constitute broader fields or sectors—that is, relatively enduring structures of global dynamics.[56] For example, International Alliance of Indigenous Peoples (IAIP) is a formal transnational nongovernmental organization that is dedicated to protecting the rights of indigenous peoples globally. Transnational coalitions composed primarily of nongovernmental organizations and grassroots groups, often including IAIP, have been formed around specific campaigns such as promoting "Latin American Indian Rights." Over time, an identifiable transnational network, focused on furthering norms on indigenous peoples and cultural preservation, has emerged with IAIP as one of its core members.[57] These different forms of transnational collective action, interacting with other types of organizations, have resulted in a transnational field on indigenous peoples. Similar transnational organizational and social fields have evolved in other issue areas such as human rights, the environment, gender, and informal workers.

The changing dynamics around big dams, as we shall see, have especially been shaped by the emergence and interaction of various forms of transnational action in the fields of indigenous peoples, human rights, and the environment, among others. The early transnational anti-dam coalitions linked nongovernmental organizations, and often drew upon preexisting transnational networks, primarily from these three issue areas. Subsequently, nongovernmental organizations were specifically established to oppose big dams and promote the wise social and environmental use of rivers globally, such as the United States based International Rivers Network (IRN). In turn, IRN participates in transnational coalitions with like-minded or strategically allied human rights, indigenous peoples, and/or environmental actors in campaigns against specific big dam projects in different parts of the world. More recently, a transnational sustainable rivers network dedicated to reforming and halting big dam building around the world has emerged—co-led by IRN, along with such groups as the Brazilian Movement of Dam Affected Peoples and the Narmada Bachao Andolan based in India (see chapter 6).

The continued transnationalization of communications, transportation, and economic exchange has clearly facilitated the emergence and perseverance of these forms of coordinated and collective action. But transnational advocacy efforts of grassroots groups, social movements, and nongovernmental organizations are most likely to be found in areas for which international and global arenas have opened up space and legitimated certain types of norms, institutions, and practices, such as the various United Nations conferences or world commissions. And to the extent that issues are already transnationalized, for example, when transnational corporations, international professional associations, multilat-eral organizations or multiple states are involved, when international laws or regimes exist, or when the issue area itself has been constructed as supranational in scope, links are not only more easily forged and maintained, they are also far more likely to be consequential.

The proliferation of various forms of transnational advocacy and contentious action generated primarily by nongovernmental organizations, grassroots groups, and social movements by itself does not directly produce visible impacts in either the international or domestic arenas. The degree to which norms and principles in issue areas, such as the environment, human rights, and indigenous peoples, among others, spread globally and become institutionalized in the procedures and structures of states, multilateral agencies, and multinational corporations, among others, greatly increases the likelihood that transnationally allied action will be effective in altering development dynamics.[58] Although the formation and consolidation of these "international regimes" or, better, "transnational fields" of norms and principles do not constitute a sufficient condition to ensure the ultimate success of transnational advocacy efforts, they are generally critical for empowering historically weak and marginal actors in world politics.[59]

The global spread and institutionalization of norms and principles have increasingly contributed to the gradual but still partial structuring of a world society.[60] In the absence of a centralized world state, various types of transnational norms—constitutive, regulatory, practical, and evaluative—have generated appropriate and acceptable identities and behaviors of actors across the world.[61] The historical period since the end of World War II has involved the dramatic spread and increasing density of these norms. In particular, deeply embedded principles of sovereignty and self-determination have historically constituted the state as a privileged form of political organization endowed with authority and agency in world society.[62] Indeed, the external legitimation of the state, its proper organizational forms and goals, represents the core of this world society.

Although it involves highly asymmetric patterns of power and influence, this partially structured world society is ostensibly constituted by formally equal states and an inter-state system. The tendency in an increasingly dense and integrated world society of formally equal states is that of diffusion and imitation. Promoting development—generally defined as progress and operationalized as increasing wealth, as well as justice operationalized in terms of increasing equity—remains one of the central and appropriate orientations of contemporary states.[63] Rapid changes in state policies and institutions often, and increasingly, reflect processes of conformity to globally spreading transnational norms of development rather than the diversity of forms and more gradual shifts expected from variations in domestic structures, interactions and processes. Similarly, in areas ranging from military organization and security, to science and education policy, to human rights, ethnicity, and environment, a remarkable degree of institutional isomorphism in the procedures and structures of states, and for that matter in various types of organizations, is evident.[64]

The area of environmental issues is an excellent example of the rise and global spread of norms that has produced international and cross-national institutional isomorphism around the world. As a one authoritative study noted:

> Widespread and mobilized world concern about the environment is heavily dependent on universalistic and scientific ideologies and principles. These have tended to arise and achieve codification in world discourse before, not after, they become local and national issues in most nation-states. And in fact, the rise of the world environmental domain clearly precedes and causes the formation of generalized national structures formalizing and managing the issues involved.[65]

The most striking evidence of the effects of this increasingly global framework is the historical process by which environmental institutions have been established by states. Prior to preparations for the United Nations Conference on Human Environment held in Stockholm in 1972, not a single state environmental bureaucracy existed. Subsequently, environmental agencies and ministries were formed at a rapid rate with approximately sixty being created by 1988, and dozens more being formed around the 1992 United Nations Conference on Environment and Development held in Rio de Janeiro, Brazil. By the late 1990s, virtually every state in the world had some type of environmental agency. Procedures such as environmental impact assessments have spread along with the formal establishment of these bureaucratic agencies, increasing the potential for development dynamics to be altered based on environmental evaluation and justification.[66]

At the international level, the United Nations Environment Program, the first environmental unit at the World Bank, which eventually became the Office of Environmental and Scientific Affairs (OESA), and even an environmental section of the International Commission on Large Dams (ICOLD), were all created after the 1972 UN Conference. The United Nations Commission on Sustainable Development, a Vice-Presidency for Environmentally Sustainable Development at the World Bank, and the Global Environmental Facility were formed as a result of the Bruntland Commission on Environment and Development and the 1992 Rio Conference that it recommended. Scores of other international organizations dedicated to various environmental issues have been established during this period. The global spread of novel environmental norms and their incorporation into the procedures and structures of states, international organizations, and multinational corporations provides strong evidence that an international environmental regime is becoming increasingly institutionalized.[67] This transnational environmental regime has been critical to the changing dynamics of big dam building, as well as broader transformation in the global political economy of development.[68]

Another clear example of the global rise and spread of norms that have contributed to altering and reconstituting the transnational political economy of development is that of human rights. Based on an analysis of the constitutions of 140 independent countries between 1870 and 1970, one study found a tremendous increase in the number of states formally ensuring a broad set of human rights. These rights were disaggregated into three categories: civil, among them, free speech and due process; political, most importantly, the vote; and social or economic, e.g., unemployment insurance and social security. In each area, there was a remarkable transformation in the number of constitutions that included these rights over this period, particularly after World War II. Civil rights expanded by 43 percent, political rights by 120 percent, while social and economic rights grew by a tremendous 340 percent.[69]

Moreover, social and economic rights appeared in virtually no constitutions before 1930, yet spread rapidly after that—especially after the United Nations Declaration of Human Rights. Although no study of newer or revised constitutions ratified since the wave of democratization that began in the 1970s is yet available, the expectation is that these documents are even more likely to include expanded lists of rights, with more on categories such as gender justice, indigenous peoples and other ethnic minorities, which have become more prominent over the past two decades.[70] By the end of the 1990s, the principle of a healthy environment was being promoted as a basic human right that should be ensured by states in their laws, procedures, and activities.[71]

At the international level, human rights norms and principles were the hallmark of the United Nations from its inception. The UN Charter—specifically the Universal Declaration of Human Rights—includes a wide-ranging set of human rights.[72] Moreover, while prior to 1948 not one international organization focusing on human rights existed, by 1990 approximately twenty-seven were significantly dedicated to this work.[73] These international organizations range from the UN Commission on Human Rights and Subcommission on the Protection of Minorities to the International Labor Organization. Numerous international conferences have been organized around human rights, and international human rights law is progressively well developed. It is thus evident that a broad international human rights regime grew extensively in the 20th century, particularly since the 1970s.[74]

The global diffusion of norms in different issue areas, however, is not a teleological or ubiquitous process marching around the world. The mechanisms by which such isomorphic institutions and change are produced should be more clearly delineated.[75] Moreover, the growing similarities in the procedures and structures of states, multilateral organizations, and multinational corporations tend to be formal and are often decoupled from actual practices.[76] In other words, institutional isomorphism does not imply complete equifinality; radical junctures between formal organization and observable practices are likely to persist.[77] This is particularly the case when the institutional incorporation of transnational norms is poorly matched or is at cross-purposes with extant and often deeply consolidated norms, principles, procedures, and structures—not to mention balances of power and interests in domestic or local contexts.[78]

Thus, theoretical and empirical analysis of the "transnational structuration" processes by which globalizing norms and principles are promoted, become institutionalized, and alter concrete practices is required.[79] The causal pathway proposed in this study involves the interaction between the global spread and institutionalization of norms interacting with the transnationally allied activities of nongovernmental organizations, peoples' groups, and social movements.[80] This latter type of transnational politics directly or indirectly contributes to the creation of novel sets of norms and principles. These norms and principles rise and spread via transnational campaigns, through world conferences and other global forums from the practices of members of professional and scientific associations, by being adopted by the most powerful states, multilateral organizations, or multinational corporations, and hence become further legitimated. For example, drafts initially formulated for the United Nations Charter barely mentioned human rights. A range of nongovernmental advocacy organizations and other actors introduced and intensively lobbied for the inclusion of human rights and were ultimately successful.[81] Subsequently, most states formally embraced the protection of these rights in their constitutions; though considerably fewer actually met the standards set by the procedures and structures they had adopted.[82] During the 1990s, more and more multinational corporations began adopting codes of conduct including human rights principles.[83] At the same time, transnational human rights networks actively began promoting "rights-based" approaches to development to contend with neo-liberal, market "price-based" models of development.

International organizations, professional associations and epistemic communities, and, critically, transnational advocacy groups, in turn, are leading propagators of globalizing norms; they author, codify, validate, and lend authoritative status to these rules of appropriate behavior. The more linked states are to this partially structured world society or, in other words, the more they interact with transnational actors, the more likely they are to incorporate transnational norms into their own institutions.[84] These formal procedures and structures can empower nongovernmental organizations, peoples' groups, and social movements by giving them new and greater institutional spaces and opportunity structures in which to gain access to and pressure states, multilateral agencies, and multinational corporations. By pressuring the powerful to be accountable to their adopted procedures and structures, these and other subaltern actors can generate changes in practices and types of outcomes that otherwise would not have occurred and seem unexplainable.

As the evidence in this study will demonstrate, the transnational institutionalization of emergent norms on human rights, as well as on indigenous peoples and the environment, has also contributed to transforming the extant international big dam regime, the dynamics of dam building more broadly and, correspondingly, the transnational political economy of development. Again, the issue area of human rights is instructive. Transnationally allied nongovernmental organizations, peoples' groups, and social movements promoting human rights across countries have utilized the increasingly pervasive international human rights regime to alter repressive state practices over the last two decades. As a result, transnational human rights coalitions and networks have contributed to processes of democratization in some of the most authoritarian regimes in Latin America, Eastern Europe, and elsewhere.[85]

Thus, the proposed constructivist analysis emphasizes the consequences of transnational norms and institutions, in addition to international military and economic power relations, for world politics. Specifically, international regimes—which have conventionally been conceived of as "sets of implicit or explicit principles, norms, rules and decision-making procedures around which actors' expectations converge"—clearly matter in ways not adequately theorized in the existing literature.[86] They may not directly constrain but create enabling conditions that increase the likelihood that the practices of states, international organizations, multinational corporations, professional and scientific associations, and other powerful actors can be more readily altered by the transnationally allied activities of nongovernmental organizations, grassroots groups, and social movements.

But many governments, in particular, remain resistant to changing their concrete practices even after the norms and principles embodied in these international regimes are formally incorporated into state structures and procedures.[87] This is partly because state institutions and governmental practices are also conditioned by domestic actors, interactions, and processes: relations between dominant classes and class coalitions, self-interested actions of political elites, inter-ethnic relations, or waves of social mobilization by subaltern groups.[88] Indeed, while transnational contentious politics interacting with the global spread and institutionalization of norms are important conditions for altering the general dynamics of development, states and other powerful actors generally have to be pressured to follow these transnational norms and principles by the mobilizing, lobbying, and monitoring activities of newly or further empowered and organized domestic and transnational actors.

REDIRECTING DEVELOPMENT: DOMESTIC DEMOCRATIZATION
AND SOCIAL MOBILIZATION

The growth of transnationally allied advocacy generated primarily by non-governmental organizations, grassroots groups, and social movements interacting with the global spread and institutionalization of norms over time has not produced identical outcomes across the world. The domestic presence of organized and sustained social mobilization as well as the presence of democratic institutions or a significant degree of democratization are critical factors that condition the broader impacts of growing transnational contentious politics and spreading global norms on the political economy of development.

More specifically, democracies that have nongovernmental organizations, peoples' groups, and social movements with the capacity to organize and mobilize large numbers of people and conduct multilevel advocacy are most likely to exhibit changes in development dynamics. Development activities—such as the building of big dams—are least likely to be altered by this form of transnational political economy, however, in states with authoritarian regimes and domestic actors that have little or no capacity to generate grassroots resistance (see figure 1.4). The latter cases are less conducive to the formation of transnational linkages among domestic and foreign actors or to the adoption of globalizing norms and principles. Because they are by definition based on the rule of law, democracies, to the contrary, increase the potential impact of globalizing norms once they have been adopted domestically.[89]

Why the levels of domestic social mobilization and advocacy might be stronger or weaker across contexts and/or over time depends on a variety of factors. Social and political activism by historically (or recently) marginalized actors was previously thought to be conditioned by two interrelated factors: the increasing scarcity and unequal distribution of resources. Development processes and outcomes, like those associated with the construction of large-scale projects such as big dams that result in environmental and social dislocation and reallocations of resources, would thus spawn public mobilization and protest. These in turn would generate or exacerbate struggles between social

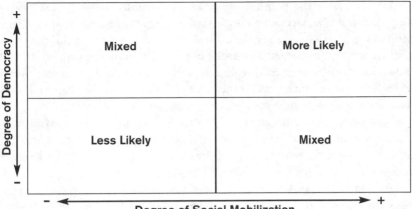

Figure 1.4. Impact of Domestic Democracy and Social Mobilization

groups and corrode extant or prevent the emergence of durable social institutions.[90]

But the organization and effectiveness of such collective political action often depend more on the mobilization of those resources that are available, the existence of facilitating social networks, and the conduciveness of the prevailing political opportunity structures.[91] These factors, moreover, are increasingly shaped by and shape the transnational linkages of domestic actors, as well as the global spread and institutionalization of norms on the environment, indigenous peoples, human rights, among others.

Significantly empowered and mobilized domestic nongovernmental organizations, grassroots groups, and social movements are crucial to the strength and viability of transnational political action "from below." These actors are critical sources of information, ideas, power, strategies, and legitimacy within transnational coalitions and networks.[92] Perhaps as importantly, they are the central actors that monitor and lobby state authorities and officials, as well as other non-state domestic and transnational actors, and hold them accountable to globalizing norms, international regimes, and domestic policies. When, due to the pressure of domestic nongovernmental organizations and peoples' groups, states incorporate the procedures and structures embodied in transnational norms, new points of access and leverage can also be opened up for these actors. Thus, for example, domestic nongovernmental organizations, grassroots groups, and social movements are greatly empowered to stop or reform big dam projects when state agencies on indigenous peoples or the environment are established.[93]

More generally, the political opportunities afforded by state institutions shape the strategies, tactics, and organizational forms of nongovernmental organizations, grassroots groups, social movements, and potentially their ultimate effectiveness.[94] With respect to impact, in particular, a fundamental feature of state institutions is the broad regime type that exists. Authoritarian regimes lower the costs and thus increase the ability for more powerful state and nonstate actors to violently repress opposing nongovernmental organizations, peoples' groups, and social movements.[95] Democratic regimes, in contrast, provide greater possibilities for domestic nongovernmental organizations and historically weaker social groups to organize, mobilize, gain access, and effectively leverage state institutions when they are engaged in contentious politics.

A number of analysts previously argued that authoritarian regimes are necessary to ensure positive environmental and development outcomes.[96] With respect to development, a central proposition was that democracy "unleashes pressures for immediate consumption, which occurs at the cost of investment, hence of growth."[97] Authoritarian regimes, however, "give political elites autonomy from distributionist pressures, they increase the government's ability to extract resources, provide public goods, and impose the short-terms costs associated with efficient economic adjustment."[98] This thesis in favor of dictatorship is partly based on a belief that a necessary, albeit not sufficient, condition for development is the existence of a developmental state, a state with a large degree of insulation against the ravages of short-run pork-barrel and rent-seeking politics, one that provides leadership in formulating coherent developmental goals and coordinates stable coalitions in favor of implementing those long-term goals.[99]

But others have proposed that democracies may just be better at producing different types of development outcomes than authoritarian regimes, such as the prevention of catastrophes (e.g., famines).[100] And more recent theoretical arguments suggest that most authoritarian rulers will prey on societies to maximize their own self-interests, and thus not promote long-term development goals as predicted. Without mechanisms that hold dictators accountable, these actors are likely to either consume scarce resources privately or squander them publicly.[101] In other words, most forms of dictatorship—and the ones that have been most likely to exist empirically—are injurious to development because they may either provide too few or too many government services.[102]

Furthermore, arguments denigrating current consumption as inimical to development were premised on the belief that investment in physical capital was most critical. Recent analyses show, however, that public investment in human capital, such as education, training, and health, which was often considered consumption in the past, is also important to generating development, even when it is defined narrowly in terms of wealth or economic growth.[103] In fact, with respect to conventional indicators of development, usually operationalized as the growth in per capita income, the most comprehensive recent research concluded that "there is no trade-off between development and democracy: democracy need not generate slower growth."[104]

By expanding the notion of development beyond the narrow sense of growth in per capita income, the case in favor of democratic regimes becomes even stronger.[105] For example, if basic civil and political liberties are included in assessments, democracies by definition are likely to be conducive to development.[106] Less tautologically, empirical studies have shown that military regimes, all of which would be classified as authoritarian, are in fact much less likely to improve the general quality of life or human development in developing countries.[107] It is also the case that population growth is far greater in authoritarian regimes than in democracies; by one estimate, the annual rate is 2.51 in the former while it is only 1.48 in the latter.[108] Higher rates of population growth can result in more resources being absorbed to maintain existing levels of per capita income, nutrition, schooling, etc.[109] Moreover, because high population growth rates have been shown to generally have negative environmental consequences,[110] democracies are much more likely to promote ecologically sustainable development.

In fact, in terms of natural resources management and environmental preservation, it was also forcefully argued in the past that the liberties and the limited government generally operating in democracies allowed individuals to overconsume, reproduce, and pollute without enforceable limits.[111] The logic was similar to that with respect to development more broadly: that strong state intervention is an absolutely essential factor in preventing environmental catastrophe, and that authoritarian regimes are much more likely to have the autonomy and capacity to execute the necessary controls. More recently, others have added that the political and economic power of industrialists—perhaps the most environmentally destructive non-state actors—often makes the process of reform in democracies extremely gradual and largely symbolic.[112]

But the worse environmental records of nondemocratic regimes, particularly those in the former Soviet Union, Eastern Europe, and China, in comparison to democratic regimes, seems to weaken the empirical support for this perspective.[113] The fact that greatest environmental gains, however limited, have been achieved in democracies contributes to skepticism about arguments that favor authoritarianism.[114] Yet, while processes of democratization seem to have been critical to improving environmental practices in the countries of Eastern Europe,[115] comparative evidence remains sparse, particularly on the experience of environmental reform in most third world countries. As we shall see, the analysis of big dam building across countries with different regime types does provide support for the hypothesis that democracies are likely to be more conducive to environmental conservation and sustainable development than actually existing nondemocratic regimes.[116]

The existence of democracy does not ensure that sustainable human development will be achieved. It is not a sufficient condition by itself. Yet there are a number of theoretical reasons to believe that these regimes will provide more supportive contexts for environmentally sustainable development practices and outcomes. A central argument of this study is that the civil rights and liberties enshrined in democratic regimes—that many scholars argued might pose dangers—actually empower domestic and transnational political action that promotes sustainable development. In particular, democratic regimes will increase the chances of impact by the activities of transnationally allied nongovernmental organizations, peoples' groups, and social movements, and the globalizing norms that promote sustainable development.

Democratic regimes are based on the premise that nonviolent collective organizing and mobilizing are acceptable and legitimate political practices. At a minimum, the range of guaranteed rights and liberties in democracies includes the right to oppose the government, freedom of expression without fear of governmental reprisal, the availability of alternative sources of information, and the liberty to support opposition parties competing to govern.[117] Governments and powerful social groups are much less able to violently repress these actors through the use of physical force because of countervailing institutions that increase the costs of doing so. Competing political parties, the press, and especially the courts not only are available to challenge repression, but also offer points of access and opportunities to leverage state institutions and alter development practices that generally do not exist in nondemocracies. These and other procedures and structures provide nongovernmental organizations, grassroots groups, and social movements room to maneuver,—room they do not have within authoritarian regimes.[118]

In democracies, nongovernmental organizations, grassroots groups, and social movements are able to more freely collect and publicize information. They can attempt to shift public opinion and lobby governmental and bureaucratic actors on problems and issues concerning development. Although powerful actors not concerned with sustainable development can also access state institutions in democracies, they can do so less secretively and monopolistically than in authoritarian regimes. Actors promoting sustainable development can monitor the activities of the former, criticize their practices, foster public debate around these issues, and even resort to protest in the streets. The importance of the free flow of information and knowledge, both within and across country borders, cannot be underestimated when it comes to actors lobbying for advances in sustainable human development. Indeed, the wide dissemination of information on the very existence of environmental and development problems, let alone knowledge about potential solutions, can also occur much more easily in democratic contexts.

Governmental and bureaucratic authorities in democracies are likely to be more accountable to proponents of sustainable development than such authorities in authoritarian regimes. Leaders in democratic regimes are generally more responsive to public opinion and pressure than leaders in nondemocratic regimes, partly because of the electoral process upon which political survival depends.[119] A free press improves considerably the ability of sustainable development advocates to shift public opinion in order to convince governments to change procedures and actions in different issue areas. Nongovernmental organizations, grassroots groups, and social movements can utilize the competition between opposing political parties to get sustainable development on the policy agenda. "Green Parties," or at the very least greener platforms, can be put forward to contest elections, and even if these parties or platforms are not successful electorally, they increase the exposure of sustainable development norms and practices.[120] Finally, governments in democracies are also likely to be more aware of transnational criticism of their sustainable development records and practices, particularly in the third world where foreign funding may suffer as a result.[121]

In democracies, nongovernmental organizations, grassroots groups, and social movements generally have another critical institutional pathway available to them, because these regimes are based on the rule of law. They can use the judicial system and courts to ensure that environmental, indigenous peoples, or human rights legislation, once it is adopted, is implemented. They can petition the courts to compel governments and bureaucratic agencies to disclose information on the destructive or repressive practices of both state and non-state actors, on the extent of environmental and human rights problems that exist, and to ensure that safeguards and protections mandated by law are followed. In democracies, matters are different because the rule of law is a central element and courts can be prime dispensers of justice. The existence and functioning of a judicial system with relatively autonomous courts can be a central countervailing institution to ensure that these acts are punished in democracies. In contrast, in nondemocratic regimes, the human rights of indigenous peoples and environmentalists are regularly abused by state authorities and non-state actors, and, in particular, by military organizations.

How democratic a government is (broadly construed) is thus likely to be a central variable that facilitates the impact of domestic mobilization on development outcomes.[122] But other aspects of states can condition the interactions and effects of transnationally allied contentious politics, globalizing norms, and domestic social mobilization and advocacy. For example, governments that are relatively autonomous in formulating policy, and that have significant capacity, can regulate the access of foreign actors and the transnational linkages that can be forged between them and domestic counterparts. Conversely, governments and bureaucracies in states that are porous find it difficult to prevent this type of access. Thus the formation of transnational coalitions and networks among domestic and foreign actors is more likely in these contexts.

After gaining access and constructing links, transnationally allied political activity may subsequently have more impact in countries that have greater state capacity to implement decisions.[123] This depends on the goals of the actors involved. If halting particular state practices, like preventing the construction of big dams, is paramount, "weak" states might actually be conducive to the effectiveness of transnational advocacy. If implementing reforms, such as progressive resettlement policies, or generating alternatives, such as water conservation, or developing nonconventional energy resources is the desired outcome, then the ability of governments in "strong" states to execute decisions will be important to longer-term effectiveness.

Nevertheless, although the strength of states may make it more or less difficult for transnationally allied political action to emerge, this factor alone rarely determines its ultimate impact.[124] In order to be effective, transnational and domestic collective action is likely to conform to the existing organization of states. For example, states vary in terms of the procedures regulating transnational and domestic actors, such as requirements for having certain types of licenses or limits on sources of external funding. This leads to variations between countries in the ways that transnational and domestic advocacy is organized and strategized.[125]

In the short run, sophisticated transnationally allied actors will strategically adapt to extant state institutions and to patterns of relations between the state and society. They may even hook up with more powerful domestic and external actors who may have convergent strategic interests. Over the longer run, however, if these actors are continually frustrated in achieving their goals, they are likely to push for broader social and institutional reforms to increase their effectiveness. In particular, as the comparison in this book of the changing dynamics of big dam building in various countries reveals, these actors are likely to pressure for greater democratization of state structures and procedures. Thus democracy is not only a condition that facilitates change in development dynamics, democratization is often a strategic means toward that change and even promoted as a goal of sustainable human development in itself. And democratization is increasingly conditioned by transnationally allied pro-democracy actors and the globalizing spread of democratic norms.[126]

DEVELOPMENT, DEMOCRACY, AND DAMS IN TRANSNATIONAL PERSPECTIVE

To summarize: in this book I argue that transnationally allied nongovernmental organizations, grassroots groups, and social movements have unexpectedly altered the political economy of development. This transformation in development has been conditioned by the global spread and international institutionalization of norms and principles in the issue-areas of environment, human rights, and indigenous peoples, among others. However, as the examination of big dam building demonstrates, these transnational structuration processes have been most successful in changing development outcomes and practices when linked to domestic actors with the ability to generate social mobilization in democratic contexts.

The two most likely competing explanations for the decline in big dam building—the decreasing availability of sites for big dams and the comparative financial and economic unviability of these projects—were addressed earlier and are further examined in the following chapters. Various implementation bottlenecks associated with ineffective decision-making structures could also have slowed down the execution of various big dam projects. Alternatively, political authorities could have reformed or stopped the building of big dams proactively, because they found more attractive forms of patronage and/or alternative means by which to satisfy the interests of powerful groups such as landed elites and industrialists. Finally, states, international organizations, and multinational or domestic companies might have independently decided that more viable options exist either for promoting development or making profits. But support for these competing hypotheses is not born out by empirical examination, given that they would require evidence that multilateral organizations, state and other non-state big-dam proponents primarily altered their big dam building policies and practices prior to, and/or separately from, the mechanisms and processes that have been posited in as central to this study.

Each of the hypotheses that as a group constitute the general theoretical argument of this book is also a potential alternative explanation. First, the lobbying of foreign and transnational nongovernmental organizations may have been the sole factor that changed the dynamics of big dam building in the third world. Second, the global spread of emergent norms on indigenous peoples, human rights, and the environment could have been so powerful that international organizations, state and non-state proponents altered their big dam building practices as a result of independently adopting and actually following these principles. Finally, it is possible that domestic advocacy and social mobilization, particularly in democracies, was sufficient by itself to stop or reform big dam projects. The diachronic and synchronic analysis that is presented in the following chapters provides support for the interactive and cumulative effects of these factors.

In order to examine the central theoretical argument of this study, as well as competing hypotheses, both cross-temporal and cross-sectional research and analytic methods have been used.[127] The cross-temporal analysis is critical for demonstrating the emergence and effects of the transnationally allied opposition to big dams as well as the global spread and international institutionalization of supportive human rights, indigenous peoples, and environmental norms over time. This specifically involves within "case," intensive process-tracing of the changing dynamics of big dam projects and big dam building in India as compared to historical trajectories in Brazil, Indonesia, China, and South Africa/Lesotho, as well as the evolution of transnationally allied anti-dam activities and their impact on the international big dam regime.[128]

The cross-sectional comparisons highlight the crucial importance of domestic social mobilization and democracy in terms of explaining outcomes, even when transnational advocacy exists and globally spreading norms have become more institutionalized. This involves structured-focused examination of the changing dynamics of big dam building in the countries of India, Brazil, Indonesia, China, and South Africa/Lesotho, as well as at the international level with respect to the World Bank and other international organizations.[129] Following the proposed theoretical argument, these country cases have been selected in order to investigate and demonstrate variation produced by different levels of domestic social mobilization and democracy (see figure 1.5).

Research for this study, which involved the gathering of both quantitative and qualitative data, is primarily based on original fieldwork conducted at numerous sites in India, Brazil, South Africa, Indonesia, China, Great Britain, the United States, and numerous other countries around the world. Statistics on big dam building, official and unofficial documents, technical reports, letters and e-mails from various government archives, international agencies, multinational corporations, and nongovernmental organizations were collected and analyzed. In addition, over three hundred interviews were conducted in English, Hindi, Gujarati, Portuguese, and Spanish with a wide range of participants, including indigenous people, peasants, landed elites, industrialists, engineers, academic experts, bureaucrats, politicians, representatives of various international organizations, as well as nongovernmental activists working in both the domestic and international arenas.[130]

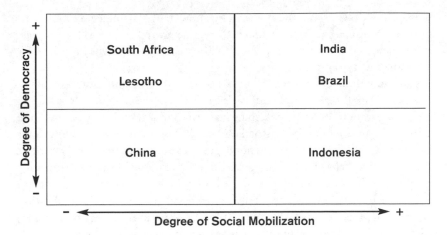

Figure 1.5. Comparative Country Case Selection

The transformation in the political economy of big dam building and development in India is examined in chapters 2 through 4. Big dam building in India is a theory generating case for this study. Big dam projects have played a prominent role in India's development since its Independence, and India was one of the leading big dam builders in the world over the second half of the 20th century. More important, India was one of the first tropical countries in which transnationally anti-dam opposition, as well as human rights, indigenous peoples, and environmental norms emerged. That country has also had the longest lasting democracy in the third world and a history of strong, domestic social mobilization. Big dam building in India is thus also a tough test of a most likely case, i.e., strong support for the theoretical argument should be found because all the variables posited as critical do operate at significant levels in this case.

In chapter 2, I examine the experience with a number of early big dam projects and the more general record of big dam building in India through the first two decades after Independence in 1947. I argue that an overwhelming state-society elite consensus under-girded support for the construction of big dams, and I show that when domestic criticism and opposition against these projects did occur, it was unsuccessful in reforming or halting them, even though a democratic political system existed in India. The striking absence of transnationally allied opposition against, and globalizing norms delegitimizing, big dams in this earlier period is empirically contrasted with the subsequent emergence of these factors in India during the 1970s. This transnationalization of contestation and newly emerging institutional context greatly strengthened domestic big dam opponents through the 1980s and 1990s, as an analysis of numerous struggles over these projects reveals. But the analysis also shows that sustained grassroots mobilization within a persistent democratic regime proved critical to the dramatically changing dynamics of big dam building and development in India over the subsequent twenty-five years.

Correspondingly, in chapters 3 and 4, I conduct an in-depth case study of the Narmada Dam Projects because of their centrality to the changing dynamics of development in India, at the international level and worldwide. Big dams were initially formulated for the Narmada River Valley by Indian authorities as far back as 1946, but construction did not begin in earnest until the early 1980s. The historical process by which the structure and functioning of India's democracy produced a three-decade delay in sanctioning the Narmada River Valley Dam Projects is depicted in chapter 3. I also show that domestic social mobilization against the proposed big dams did not exist during most of this period and, when it finally did emerge in the late 1970s, was ineffective in independently altering the course of projects proposed for the Narmada River. It was only after a transnational coalition of nongovernmental organizations and grassroots tribal peoples' groups was forged, conducting a five-year campaign to defend the human rights of tribal peoples to be displaced by the Sardar Sarovar major dam, that proponents, such as Indian and World Bank authorities, reformed their policies and practices on resettlement and rehabilitation.

In chapter 4, I continue my ethnographic and process-tracing examination of the trajectory of Narmada River Valley Dam Projects from the early 1980s on, focusing on the formation of a second transnational coalition dedicated to halting the Sardar Sarovar major dam altogether and thereby altering the course of the broader initiative. I demonstrate that the continuing and intensifying transnationalization of contestation and the changing institutional context for big dam building—particularly the rise, spread, and deepening of environmentalism, human rights principles, and norms on indigenous peoples globally and domestically in India—were critical to the eventual stalling of construction on these projects. Lobbying by foreign and/or transnational nongovernmental organizations would not have produced this outcome independently, however, had it not been linked to strong, domestic social mobilization. And once again, India's democratic regime strongly conditioned the effects of the transnationally allied opposition and domestically institutionalized global norms.

In order to further refine my general theoretical argument and compare it with alternatives, I conduct a comparative historical cross-country analysis in chapter 5. In particular, I examine the emergence and effectiveness of anti-dam activity in four prolific, big dam building, third world contexts: Brazil, Indonesia, South Africa/Lesotho, and China. All these cases had similar, relatively robust, records of big dam building to that of India from the 1950s through the 1970s.

As in the case of India, it is expected that the blocking and/or reform of these projects in Brazil noticeably increased because of the formation of transnationally allied opposition, the increasing domestic institutionalization of global norms on the environment, human rights, and indigenous peoples, the growth and increasing organization of domestic anti-dam mobilization as well as the democratization of the political regime since the late 1970s. The dynamics of big dam building in South Africa and Lesotho, however, demonstrate that the impact of primarily external transnational organizing, the existence of supportive globalizing norms, and even domestic democratization during the 1990s will be limited without the presence of strong, grassroots social mobilization and advocacy.

In Indonesia, likewise, big dam projects have by and large continued to be executed. Although the existence of lobbying by domestic nongovernmental organizations and grassroots anti-dam mobilization has resulted in some reforms, the authoritarian regime that persisted in that country until recently has generally given big dam proponents a relatively unchecked ability to repress opponents. Even with the adoption and internalization of environmental norms and principles into the Indonesian state, the subversion of human rights, judicial, and other democratic procedures has limited the possibilities for effective transnational links to be formed and for domestic opposition to be effective.

In the case of China, the absence of both domestic social mobilization and the persistence of a dominant authoritarian regime has resulted in even less change in the dynamics of big dam building than in Indonesia. For example, despite the fact that an emergent transnational anti-dam network has prevented foreign donors (e.g., the United States Export Import Bank) and multilateral development agencies (e.g., the World Bank) from supporting China in its endeavor to build the mammoth Three Gorges Project, construction has continued. But the case of the Three Gorges does highlight that even when transnationally allied anti-dam activity has encountered unfavorable domestic-level conditions, it has increasingly been able to have a noticeable and non-negligible effect on development processes and outcomes.

In the sixth and final chapter of this book, I correspondingly investigate the formation and institutionalization of a transnational anti-dam network and its effects on big dam building dynamics, particularly at the international level with respect to the policies and practices of the World Bank and other powerful pro-dam actors. I also examine two critical transnational gatherings that took place in the late 1990s. The first conference contributed to the building of a transnational anti-dam network by bringing together critics and dam-affected peoples from twenty different countries. Leading members of this emergent transnational network, along with representatives of multinational corporations, international agencies, professional associations, states, and others participated in a second and subsequent conference that established a novel global governance initiative to review the historical performance of these projects and establish new global big dam building norms and principles for the 21st century. Given these events, I propose that the "international big dam regime" or "transnational big dam field" was dramatically altered during the 20th century.

In closing, I reevaluate the theoretical explanation and the alternatives offered to account for this unexpected change in big dam building in light of the weight of evidence offered in the book. After some empirical and conceptual extensions, I suggest that a profound transformation in the transnational political economy of development has occurred, reflected by the dramatic fall and reforms in big dam building and shifts in the international big dam regime. I conclude with some reflections on broader empirical and theoretical implications of the study, as well as identify avenues for future research. These include a proposed constructivist transnationalism agenda for the social sciences as well as a linked scholarly agenda on development dynamics in a transnationalized world.

❖

CHAPTER TWO

Dams, Democracy, and Development in India

When I walked around the site, I thought these days, the biggest temple and mosque and gurdwara is the place where man works for the good of mankind. Which place can be greater than the Bhakra Nangal Project, where thousands of men have worked or shed their blood and sweat and laid down their lives as well? Where can be a holier place than this, which can we regard as higher?

JAWAHARLAL NEHRU, India's first Prime Minister.[1]

Silent Valley, Ichampalli, Koel Karo, Tehri, Narmada; these are now familiar names. Along with being the names of big dam projects, they have become synonymous with peoples' struggles. They used to be heralded as temples of our country's march towards progress. Now, these harbingers of development are seen as tombs of destruction.

BABA AMTE, veteran Gandhian social activist and renown spiritual leader.[2]

BIG DAMS IN INDIA: TEMPLES OR TOMBS?[3]

The changing dynamics of big dam building and development in India both reflect and contributed to the broader transformation in the transnational political economy of dam building and development during the 20th century. A top-down, state-led, economic-growth focused and technocratic development vision, in which big dam projects played a central role, was adopted by Indian authorities at Independence. This development vision that helped legitimate and naturalize the building of big dams was more or less the same as that adopted by most states across the world (see chapter 5), and supported by international development agencies like the World Bank (see chapter 6) after World War II.

There were episodes of domestic criticism against this development approach more generally and against big dams in particular during the 1950s and 1960s in India. But this contestation did not tangibly alter these projects or the broader development vision for which they were central activities and symbols, even though domestic democratic institutions existed. In fact, had it not been for the decreasing ability of the Indian state to complete the increasing number of big dams that were planned during these first decades after Independence, many of the sites available for these projects might have been utilized by 1970.

But domestic criticism and opposition grew and spread dramatically from the 1970s on in India, often against projects that had been initiated decades earlier but never completed, or against others that had been resurrected after lying dormant for long periods of time. By the 1980s, practically every big dam across India faced some form of organized resistance.[4] This domestic opposition was progressively empowered during this period by the gradual institutionalization of globalizing norms on the environment, indigenous peoples, and human rights domestically in India, interacting with links constructed with rapidly proliferating transnationally allied dam critics.

As a result, conflicts between proponents who viewed big dams as temples and opponents who saw them as tombs reached unprecedented levels, and a dramatic slowdown in the construction of these projects emerged in the 1980s and 1990s.[5] This outcome was greatly facilitated by the presence of democratic institutions in India that provided critical points of access and leverage to, and circumscribed the ability of big dam proponents to repress opponents. And these intensified transnational struggles around big dams were indicative of changes in development dynamics in India, across the third world, and at the international level.

THE GROWTH AND STAGNATION OF BIG DAM BUILDING IN INDIA

Coordinated development through big dam building on many of India's waterways was originally investigated and planned by British authorities during the colonial period. The historical examination of India's Narmada Projects in the next chapter vividly demonstrates this pattern. Although those pre-independence river valley development projects that were built by imperial authorities consisted primarily of weirs and smaller dams, more than two hundred big dams taller than fifteen meters in height had been constructed in India by 1950. A technical and bureaucratic foundation was thus already established for the massive big dam building development drive subsequently initiated by postcolonial Indian authorities.[6]

At the time of Independence in 1947, India's governing coalition—led at the federal level by the nationalist hero Prime Minister Jawaharlal Nehru, a practically unchallenged Congress Party, and a "steel frame" bureaucracy—was faced with weak regional State capacities and a relatively demobilized civil society. As a result, Indian political and bureaucratic authorities were able to execute development strategy, policies, and projects quite autonomously and effectively.[7] The vision of development adopted was a progress-as-economic-growth-oriented one, and capital-intensive initiatives and the exploitation of natural resources were not only promoted, they were prioritized.[8] The selection of this approach entailed the marginalization of a more indigenously rooted vision of development defined in terms of bottom-up participatory processes and social justice outcomes often associated with the views of Mahatma Gandhi (who was assassinated in 1948).[9]

Big dam building, especially multipurpose river valley projects that would generate irrigation and power, and provide flood control became a centerpiece of this development vision.[10] Given these ostensible functional benefits, the espoused development vision, and the fact that these projects satisfied the interests of what has been termed India's "dominant coalition of proprietary classes": irrigation to rich farmers, electricity to industrialists, and good-paying and prestigious work to skilled professionals (e.g., civil servants and engineers), the overwhelming political consensus behind big dam building in India during the 1950s and 1960s was not particularly surprising.[11] Correspondingly, as one would expect, political authorities were able to use these projects as sources of control and patronage in India's newly established democratic system.[12] Big dam and river valley development projects were also being financially and technically supported by foreign donors including the United States, the Soviet Union, and the World Bank, among others.[13]

Great numbers of big dams were initiated right at the very start of the post-Independence era. Public spending in India's first Five Year Plan (1952–57) was dominated by three major river valley projects: Bhakra Nangal, Damodar Valley, and Hirakud. The joint decision by the federal Government's Ministry of Irrigation and Power, Ministry of Finance, and Planning Commission to have the Central Water and Power Commission build the Bhakra-Nangal Project primarily with Indian personnel inaugurated this developmentally driven big dam building era in India. This gigantic 740-foot-high structure provided specialized training for Indian engineers in all aspects of formulating and implementing a big dam project. Moreover, it was at the commissioning of the Bhakra Nangal Project that Prime Minister Nehru described big dams as the "modern temples of India."[14] The Damodar Valley Project, as mentioned in chapter 1, was explicitly modeled after the Tennessee Valley Authority in the United States.

The case of Hirakud, the other one of Independent India's first big, multi-purpose projects, provides a clear demonstration of the inability of domestic opposition to alter big dam building in India during this initial period. It shows the purely domestic nature of the resistance, the lack of environmental norms guiding practices, and the absence of nongovernmental organizations lobbying to protect the rights of the peoples to be negatively affected. Without the support of transnationally allied advocacy and the legitimation provided by globalizing norms on human rights, indigenous peoples, and the environment, domestic big dam opponents were too politically weak to successfully utilize the procedures and institutions of India's democratic regime to either halt or even reform the Hirakud Project.

The nature of the domestic opposition to the Hirakud Dam is strikingly similar to domestic anti-dam struggles that emerged in greater numbers in India during the 1970s (see next section). Following the inauguration of work on the multipurpose project on the Mahanadi River in 1946 (the year before Independence), a series of public meetings were coordinated by Congress Party leaders in the State of Orissa to discuss its social and economic costs.[15] The people to be negatively affected by the project decided to demand compensation (for the loss of lands that would be submerged) and the rehabilitation of villages as collective units in new locations to be installed with basic facilities, such as water, electricity, roads, and schools. When the federal Government's Minister of Irrigation and Power visited the area, four thousand people protested to publicize these demands.

The large public demonstrations did not deter authorities from moving forward with the project. In response, a series of strikes was organized by local groups and a resolution was adopted to promote the separation of Sambalpur District (the area in which 249 villages were to be submerged by the Hirakud Dam) from the State of Orissa. The demand for secession, while forming an important mobilizational strategy among the villagers, resulted in a split in the broader opposition to the project. A number of local Congress Party leaders withdrew from the agitation when it became clear that the resolution had been strongly disapproved of by Orissa State and federal Congress Party leaders, including Mahatma Gandhi. Given the recent partition of British India into Independent India and Independent Pakistan, and the violence this had generated, any proposals for secession had become unacceptable.[16]

But the grassroots mobilization only increased in strength. Disobeying orders prohibiting further public opposition, the remaining local anti-dam leaders took to the streets with thirty thousand villagers under the banner of the Communist Party in the soon to be submerged areas of Sambalpur and Padampur, rallying in front of the Governor of Orissa. As a Government of Orissa State report later stated, "The people had great doubts regarding the benefits (of the Hirakud Project) in comparison to the present loss of their ancestral homes and best cultivated land."[17] When clashes broke out between the anti-dam marchers and pro-dam construction workers, government officials quickly declared the agitation illegal, dispersed the demonstrators with a police lathi-cane (police baton) charge, and arrested the protest leaders.

With little power to alter the growing political support behind the project that went as high as the Prime Minister himself, and weakened by the arrest and later cooption of key leaders, the anti-Hirakud campaign soon came to halt. Of the 249 villages submerged by the project, resulting in the displacement of over 20,000 people, only 33 were relocated.[18] Signaling that the displacement of people from the construction of big dams in the name of development would become a common feature in post-Independence India, Prime Minister Jawaharlal Nehru told villagers while laying the foundation stone for the project on April 12,1948, "If you are to suffer, you should suffer in the interest of the country."[19] Before his death in 1964, Nehru himself inaugurated construction of many of the biggest dam projects in India, including the predecessor to the Sardar Sarovar major dam component of the Narmada Projects, examined in the next two chapters of this book.[20]

Due to the overwhelming ideological consensus, powerful coalition of actors, and relative autonomy and capacity of the state in support of these projects, India successfully became one of the top five big-dam-building countries in the world by the 1970s. Nearly 900 big dams were completed between 1951 and 1982 (see table 2.1). To put this success in comparative perspective, only China, the United States, the former Soviet Union, and Japan had built more big dams than India after World War II.[21]

But delays in the formulation and implementation of these projects also became increasingly common in India during this period. For example, in one sur-

Table 2.1. Big Dams in India

Completed height (meters)	Prior to 1950	1951–1982
15–30	171	710
30–99	31	163
100–149	0	8
over 150	0	2
Total	202	883

Source: Compiled from Central Board of Irrigation and Power, Large Dams in India, vol. 1, no. 197 (New Delhi: Central Board of Irrigation and Power, October 1987), 205.

vey of 192 major river valley projects begun from 1950 to 1980, only 42 were found to have been completed as of 1984.[22] Focusing on this lack of project execution, India's Eighth Five Year Plan argued that the problem, stemming right from the First Plan soon after Independence:

> has been the continued tendency to start more and more new projects resulting in wanton proliferation of projects. . . . Though all the Plans, without exception, declared their intention to give priority to complete the ongoing schemes, the addition of new schemes continued unabated.[23]

Very few sites would have remained un-dammed by the 1970s had all the planned projects initiated during the first twenty years after Independence been completed on schedule.

A central cause of India's inability to implement the increasing number of big dam projects initiated was the complex political and administrative machinery that evolved during this period.[24] For example, although water was not officially made a concurrent subject to be coordinated by the Centre and regional states, a large degree of jurisdictional overlap emerged with respect to the federal distribution of authority in this area.[25] According to Entry 17 of the State List in the Indian Constitution, "Water, that is to say, water supplies, irrigation and canals, drainage and embankments, water storage and water power," was to be controlled by (regional) States in India. But this entry in the State List is "subject to the provisions of Entry 56 of the Union List," which charges the federal government with "Regulation and development of inter-State rivers and river valleys to the extent to which such regulation and development under the control of the Union is declared by Parliament by law to be expedient in the public interest." These provisions clearly gave the Centre the authority to intervene in water resources management and development.

The distribution of authority between the Centre and the States over water resources, and thus for big dam building, was even more convoluted in practice because almost all the major rivers in India flow through more that one State.[26] In recognition of this fact, Article 262 of the Indian Constitution grants further authority to the Centre over water issues. "Parliament may by law provide for the adjudication of any dispute or complaint with respect to the use, distribution or control of waters of, or in, any inter-State river or river valley," and, "Parliament may by law provide that neither the Supreme Court nor any other court shall exercise jurisdiction in respect of any such dispute or complaint. . . ."[27] The second clause also highlights the tension between parliamentary sovereignty and judicial review that exists within India's democracy, a factor that became increasingly important to the dynamics of big dam building, especially from mid-1970s on (see chapters 3 and 4).

These constitutional provisions were utilized by Parliament to subsequently enact two pieces of legislation in 1956 that further complicated the distribution of control between the Centre and States with respect to water resource management and big dam building in India. The "River Boards Act" authorized the appointment of river valley authorities by the Centre in consultation with the States to advise on integrated development of inter-state rivers and river basins.[28] The "Inter-State River Water Disputes Act," to the contrary, gave States the right to petition the Centre to establish a water disputes tribunal when their interests in the waters of an inter-state river were affected prejudicially by the actions of one of more other states.[29] The latter legislation was invoked three times since Independence, and was a central component in the trajectory of the Narmada Projects (see chapters 3 and 4), partially as a result of the difficulties that have been encountered in establishing cooperative river basin arrangements among States under the "River Boards Act."[30]

The Centre's role in water resources development was further expanded relative to the States by two other persistent features of India's political system. First, entry 20 in the Concurrent List of the Constitution mandates coordinated Centre-State economic and social planning for India's development. By virtue of this provision, major and medium irrigation, hydropower, flood-control, and multipurpose projects had to be cleared by the Planning Commission at the federal level from the first Five Year Plan onwards. Second, given the financial and technical dependence of the States on the Centre, States thus also had to acquire additional clearances from the Finance Ministry and the designated federal agencies in charge of water and power when proposing big dam projects within their territories.[31]

These requirements on regional States to acquire technical, financial, and planning clearances from the Centre reflect the complex bureaucratic framework that progressively slowed big dam building and water resources development in India. For example, the Central Water and Power Commission (CWPC) was established in 1951 by the federal Ministry of Natural Resources and was "charged with the general responsibility of initiating, co-ordinating and furthering, in consultation with the State Governments concerned, schemes for the control, conservation and utilization of water resources," and to, "if so required also undertake the construction and execution of such schemes."[32] The CWPC was formed by combining two colonial agencies, the Central Electricity Commission and the Central Water, Irrigation and Navigation Commission, precisely in order to unify and integrate water and power development in India. The CWPC was then incorporated into Ministry of Irrigation and Power, itself created in 1952.

Thus, big dam projects have been dependent on the initiative and approval of multiple bureaucratic agencies at the federal level: the Central Water and

Power Commission (or CWC and CPC); the Ministry of Irrigation and Power (or the Ministry of Water Resources and the Ministry of Power); the Finance Ministry and the Planning Commission. At various points of political and administrative reorganization in India's history, there has been a separation between the Ministry of Irrigation (or Water Resources) and the Ministry of Power (or Energy), as well as a corresponding division between the Central Water and Central Power Commissions (CWC and CPC). Moreover, similar procedures and organizational structures exist at the State level, further complicating the process of big dam building and water resources development in India. And when different political parties have controlled the State(s) versus the Centre in India's federal democratic system, the potential for conflicts resulting in delays in the completion of these projects greatly increased.[33]

Despite this increasingly complex institutional maze in India's water resources sector, many big dams were completed at the same time the total number of these projects under construction continued to grow. Thus the total number of big dams built in India stood at between eleven and twelve hundred by the 1990s which represents a nearly 600 percent increase since Independence.[34] Moreover, most of these big dams, approximately 85 percent, were completed from the 1950s through the 1970s (see table 2.1). By official figures, India had built 797 big dams between 1951 and 1977 at a rate of approximately 30 per year. From 1978 to 1982, 86 additional big dams were completed, while between 1983 and 1986 only 35 more big dams were built.

However, as of 1990, there were 182 ongoing major irrigation projects, according to the final version of India's Eighth Five Year Plan (1992–1997).[35] Consequently, by one estimate, more than 550 big dams were still at various stages of formulation or implementation in India at the beginning of the 1990s.[36] The corresponding annual completion rate for big dams had correspondingly fallen to 7 by the mid-1980s or by 75 percent within a decade. A variety of sources indicate that this figure either remained stagnant or continued to decline through the 1990s.[37]

THE CHANGING INSTITUTIONAL CONTEXT WITH RESPECT
TO THE ENVIRONMENT IN INDIA

Neither the absence of suitable sites nor the inefficiency of the Indian political and administrative framework can fully account for the rapid deceleration in the rate of completing big dam projects from the 1970s on. As noted above, the number of big dams proposed or under construction remained high and perhaps even increased during that decade. Most of the multiple and overlapping procedures and organizational structures that caused management and execution failures had been established well before this period.[38] Governmental budgets for big dams continued to increase in absolute terms, so financing for these projects was available, as well.[39] Thus, other causes for the decline in big dam building must be found.

The dramatic decline in big dam building in India during the last quarter of the 20th century was increasingly conditioned by the mounting transnationally allied opposition against these projects. But without the progressive domestic institutionalization of global norms on the environment, indigenous peoples, and human rights denaturalizing big dams, this transformation would not have occurred. In 1970, for example, four thousand people to be adversely affected by the Pong Dam in the northern State of Himachal Pradesh protested because adequate resettlement plans had not been formulated.[40] Although the demonstrators stopped work for fifteen days, the dam was eventually completed.[41] In contrast, numerous struggles against big dams—from the Silent Valley to the Bodhghat to the Subarnarekha to the Tehri, or in favor of the rights of those displaced in cases where these projects had already been constructed—were initiated. These became increasingly effective over the subsequent two decades.[42]

Emergent environmental consciousness and institutions in India, which partly responded and contributed to the global spread of novel environmental norms, were pivotal to the transformation of big dam building in India after the 1970s. The United Nations Conference on Environment, held in Stockholm in 1972, had put environmental principles squarely on the international agenda.[43] In India, the process of integrating environmental factors in development policy formulation had begun slightly earlier with the Fourth Five Year Plan (1969–1974), which was actually completed in 1970. In the official plan document, Indian authorities formally stated for the first time that "planning for harmonious development . . . is possible only on the basis of a comprehensive appraisal of environmental issues."[44]

But no concrete changes in actual development institutions or practices consistent with this principle were proposed at the time. It was only after the further legitimation provided by the Stockholm Conference that the federal Government of India translated these emergent norms and principles into specific domestic procedures and organizational structures. A high-level "Committee on Human Environment" was established by the Centre in 1972, because the UN Conference organizers had asked member states to prepare country reports. The committee's report led to the creation of the National Committee on Environmental Planning and Coordination (NCEPC) within the prestigious Department of Science and Technology in 1974 to serve as an advisory body on environmental issues.[45] Chaired by Prime Minster Indira Gandhi herself, who had participated in the Stockholm Conference, the NCEPC organized various task forces to deal with issues, such as identifying environmental problems, carrying out investigations, and integrating environmental concerns into specific development programs and projects.

The functional importance of building an environmental protection program in the country had not been established and civil society mobilization focused on ecological issues was just emerging in India in the early 1970s. Thus, the formal institutional changes within the Indian state with respect to the environment were powerfully motivated by progressively spreading global norms and principles. For example, India's Parliament drafted "The Prevention of Air Pollution Act" in 1972 and passed "The Water (Prevention and Control of Pollution) Act" in 1974 with few domestic scientific studies documenting problems in these areas. Moreover, civil society groups lobbying for environmental action were weak in India at the time.

The newly adopted environmental procedures and organizational structures altered the institutional context for big dam building in India. In 1977, based on recommendations from the NCEPC, Prime Minister Indira Gandhi issued a directive that environmental impact assessments be completed by federal agencies for all medium and major irrigation projects—that included all big dam projects. Within three years, based on the recommendations of the high level, federal government appointed Tiwari Committee on the state of India's environment, the Prime Minister also sanctioned the creation of a federal-level Department of Environment, which later became a Ministry of Environment and Forests in 1985, and Parliament passed the Forest Conservation Act.[46] Regional States were mandated to secure environmental and forest clearances in addition to the technical, financial, and planning sanctions previously required.

THE TRANSNATIONAL CAMPAIGN AGAINST THE SILENT VALLEY PROJECT

These institutional changes, which quickly filtered down to the State level, were critical to the changing dynamics of big dam building in India, but other factors including the rise of transnationally allied advocacy and increasing role of domestic anti-dam mobilization also contributed. The first successful transnational campaign opposing a big dam project in India, and perhaps the world, was waged against the Silent Valley Project that was to be built in the southern Indian State of Kerala. Around the same time, initial domestic resistance to the building of the gigantic Tehri Dam at the foot of the Himalayas in the northern Indian State of Uttar Pradesh had begun, a struggle that under various incarnations continued into the 1990s. These two cases, as well as numbers of others, demonstrate the growth of grassroots mobilization, the forging of transnational linkages among rapidly proliferating big dam opponents from all around the world, and increasing domestic institutionalization of globalizing norms legitimizing critiques of big dams that occurred in India from the 1970s on. They also show the crucial importance of democratic institutions to the success of these campaigns.

The Silent Valley Project, like many others in India, was initiated prior to Independence. British Colonial officials identified a site in the Silent Valley as ideal for the generation of hydropower as early as 1929. The Silent Valley occupies an area of approximately 8950 hectares in southwest India, primarily in the State of Kerala. Combined with the Nilgiri and Nilambur forests to the north and the Attapppadi forests to the east, the Silent Valley comprises part of 40,000 hectares of forest in the broader ecosystem known as India's Western Ghats. The Kunthi River originates at a height of more than 7,000 feet on the outer run of the Nilgiri forests, descends rapidly to approximately 3,500 feet on the northern edge of the plateau, and then pursues a southward course before cascading down to the Mannarghat plains through a gorge at an elevation of over 3,000 feet. This gorge was the site selected for the Silent Valley Project.[47]

No criticism against the project was raised at the time when the Kerala State Electricity Board (KSEB) published a legally required notification in the *Gazette of India* in June 1973 regarding the commencement of the more than 390-feet-high, 120-megawatt Silent Valley hydropower dam. Post-colonial Indian authorities only began conducting detailed technical investigations based on previous colonial studies in 1958. After delays that lasted 15 years associated with the institutional and bureaucratic difficulties of sanctioning big dam projects in India discussed in the previous section, the Silent Valley Hydro-Electric Project was finally cleared by India's Planning Commission. Work on the project itself was subsequently stalled, because of a shortage of funds until 1976 when the KSEB decided to resume its efforts.

The conflict over the Silent Valley both contributed to and was conditioned by the growth of environmentalism in India. Nalini Jayal was the Joint Secretary in the federal Government's Ministry of Agriculture in charge of wildlife and forests in 1976, and later became a Joint Secretary in the Department of Environment. In these positions, Jayal reviewed the initial Kerala State proposal for clearance of forests in the Silent Valley and alerted various leaders in the federal Government to the environmental dangers associated with the construction of the big dam.[48]

Transnational criticism of the project began almost immediately after the project was re-initiated. Zafar Futehally, Vice-President of the Indian chapter of the World Wildlife Fund—a transnational environmental nongovernmental organization launched domestically in 1969—had become concerned with the fate of the Western Ghats. After discussing the destructive impacts of the Silent Valley Project on the Western Ghats in a personal interview with Prime Minister Indira Gandhi, he sent a memo on the topic to her. The critique was referred to Ashok Kosla, a scientist working in the newly formed NCEPC. Khosla reviewed the memo and returned it to the Prime Minister with his own recommendations. Gandhi responded by creating a task force within the NCEPC to investigate the environmental problems of the Western Ghats, and the Silent Valley Project in particular, with Futehally as chair in April of 1976.

Grassroots mobilization against the Silent Valley Project emerged at the same time in 1976. A local group of teachers and scientists in Kerala, belonging to the Kerala Sastra Sahitya Parishat (KSSP), became aware of the negative social and environmental consequences of the Silent Valley Project. However, aside from writing a few letters to government officials on the destruction of forests resulting from construction of the hydroelectric dam, the KSSP did not announce its opposition until two years later, when a mass signature campaign was initiated. At that time, a memorandum protesting construction signed by 600 prominent citizens, academics, and students was sent to the Chief Minister, A. K. Anthony, and a call for a halt to construction on the project was passed at the KSSP's annual conference.

The NCEPC's critical "Report of the Task Force for the Ecological Planning of the Western Ghats" was finalized during the same period. It stated that the outcome of the Silent Valley Project, "will be that the last vestige of natural climax vegetation of the region, and one of the last remaining in the country, will be lost to posterity and various adverse ecological consequences listed in this report will follow," and concluded that, "the project should be abandoned and the area declared a biosphere reserve." But the NCEPC task force also proposed 17 safeguard measures in case federal and State authorities deemed that the big dam had to be built, despite the objections raised against it in the report.[49]

In response to the NCEPC report and the mounting grassroots resistance, the Kerala Assembly unanimously passed a resolution to hasten the construction of the dam. An all-party delegation of State assembly-persons met then Prime Minister Morarji Desai to further demonstrate their resolve to build the project. Desai suggested that if the Kerala Government would promulgate an act giving legal sanction to the NCEPC's recommended safeguards, he would not obstruct construction of the project. A temporary ordinance to protect the ecological balance of the Silent Valley was issued by the Kerala State Minister soon thereafter. In March of 1979, the Kerala assembly passed a second act that enshrined these same principles, in addition to constituting a monitoring committee to ensure that the safeguards were implemented.

Simultaneously, the controversy over the Silent Valley Project was becoming transnationalized. At the urging of Keralite J. C. Daniel of the Bombay Natural Historical Society (one of the oldest conservation nongovernmental organizations in India), a resolution appealing to the Government of India to more effectively preserve the Silent Valley was passed at a conference of the Switzerland-based International Union for the Conservation of Nature (IUCN), one of the oldest transnational environmental organizations in the world.[50] The IUCN along with the World Wildlife Fund (WWF), and other nongovernmental organizations, highlighted internationally the issue of the Silent Valley Project as part of a growing transnational campaign to save rain forests around the world. This generated considerable political pressure on Indian authorities.[51] In addition, these actors provided some critical information, such as aerial photography and satellite imagery of the Silent Valley Forests, which Zafar Futehally and other critics put to strategic use domestically in India.

By then the grassroots mobilization was also beginning to have an impact. The KSSP discussed the Silent Valley Project in every available public arena, including in numerous rural science forums which it organized across the State, and published anti-dam articles both in the regional newspapers and journals. A small group of leading KSSP members also conducted the first comprehensive, nongovernmental assessment of the project. In the document, the authors argued that the project would contribute to the destruction of the Silent Valley forests, was not justified because forty percent of Kerala's power was already being exported to other States, and that the irrigation to be generated could be delivered much more cost effectively by utilizing the ground water potential in the area.[52] Another local group fighting to protect the Silent Valley forests led by Joseph John filed a writ petition in the Kerala High Court against the project in April 1979. The judge, after the hearing was held in August, ordered a two-week stay on construction and admitted another petition against the project filed by the KSSP into court.

The Silent Valley issue was increasingly becoming a political issue all across India and soon spread its way through India's democratic system. In January 1979, the federal Government of India's Finance Minister, H. M. Patel, acting as President of the Indian Wildlife Board, urged Kerala State officials to halt the project. In May, thirty ministers of India's federal parliament sent a similar appeal. By the summer, individuals and organizations outside Kerala, particularly well-known conservationists from Bombay, formed the Save Silent Valley Committee. The members of the Save Silent Valley Committee, and particularly Dilnavaz Variava who was an honorary advisor to the WWF, proved to be critical actors in coordinating the anti-dam campaign. They often did this simply by passing information between the local anti-dam groups and sympathetic federal government officials in Delhi, such as Nalini Jayal of the NCEPC.

The Silent Valley Project was also driving a wedge within the Indian state. Battle lines were drawn between bureaucratic agencies: the Department of Science and Technology and the Ministries of Irrigation and Power supported the project; the NCEPC and the Ministry of Agriculture—which still oversaw the issue of forests at the time—opposed it. The NCEPC, in fact, had by then already decided to declare its opposition to the Silent Valley hydroelectric dam and to persuade Kerala State authorities to collaboratively find alternatives to it. But the head of the federal Government's prestigious Department of Science and Technology stated that the environmentalists did not have a strong enough case to preserve the Silent Valley, and the Ministry of Power expressed that it could not independently halt a project on ecological considerations.

The Silent Valley Project also bled into democratic electoral and party politics. Parliamentary elections were announced for January 1980 after the collapse of the Janata-coalition government, led by Prime Minister Morarji Desai, at the federal level in July of 1979. Seizing the opportunity of a lame duck administration, the members of the Save Silent Valley Committee lobbied interim Prime Minister Charan Singh to take personal action in October. Singh had received a confidential background note from the NCEPC, which was critical of the project, and promised that he would look into the matter. By that point, members of the Save Silent Valley Committee had already convinced most members of the NCEPC task force on the Western Ghats, including Chairman Futehally, to refuse to serve on the monitoring committee constituted by the Kerala State Electricity Board, which was to implement the safeguards proposed in the task force's original report.

A second, officially sanctioned, federal level government study on the Silent Valley was conducted by the Secretary of India's Agricultural Ministry in response to the public criticism.[53] The Agricultural Ministry at this point was still in charge of forest management and development in India. The Ministry of Agriculture's report concluded that the Silent Valley Forest indeed had to be preserved and that the way to do this was to formulate alternative projects that would meet the irrigation and power needs of Kerala. Persuaded by the personal lobbying of the Secretary of Agriculture, and the growing anti-dam mobilization, interim Prime Minister Charan Singh sent a letter to the Kerala Chief Minister asking him to drop the project because of the growing controversy.

The Save Silent Valley Committee further found a receptive audience with Indira Gandhi who, not to be outdone by opposing political parties, requested her Congress(I) Party faithful in Kerala to withdraw their support from the project. Gandhi's action was a result of two converging motivations. First, after being ousted from power in 1977, she was mounting a populist, electoral campaign to return to the Prime Minister's office. Supporting the broad-based opposition to the Silent Valley Project was thus partly a strategic, electoral tactic. Secondly, she had become increasingly more attuned to environmental issues compared with other Indian politicians since her involvement in the Stockholm UN Environment Conference of 1972.

In addition to the courts and political parties, the press—another critical institution in India's democratic system—began to play a progressively greater role in the dynamics surrounding the Silent Valley Project at this time. An example of the importance of the media occurred in October of 1979. Arguing that any public opposition would be sub-judice while the High Court hearing on the original petition was still being conducted, Kerala officials obtained an injunction preventing an open meeting that had been organized by the Save Silent Valley Committee for October 1979 in the State capital of Trivandarum.[54] *The Hindu* quickly and unequivocally printed a condemnation of the tactics of the Kerala State authorities as preventing the freedom of expression and organization on which India's democracy rested. *The Hindu* had been the leading national newspaper to support the anti-dam campaign and greatly contributed to its gaining country-wide attention. Numerous critical editorials, full-page features, and letters were published on the Silent Valley Project in *The Hindu,* as well as in other regional and national papers. As a result, the controversy became even more publicized, adding further momentum to the anti-dam campaign.

The conflict escalated further when Kerala authorities hardened their resolve to implement the project. They were especially encouraged when the Kerala High Court dismissed the two writ petitions filed against the project and lifted the stay on construction in early January 1980.[55] The Kerala State Electricity Board, in particular, moved quickly, trying to complete as much construction work as possible.

But Indira Gandhi's re-election to the office of Prime Minister that very same month gave the anti-dam proponents an opportunity to once again lobby for federal intervention to halt the project. Members of the Save Silent Valley Committee sent letters and cables to Gandhi after her re-election. They congratulated her and commended her personal interest in the Silent Valley, stating that only with the creation of a federal Department of Environment and her direct involvement would the controversy be resolved. IUCN Director General David Munro also sent Gandhi a letter from Switzerland asking her to take steps to stop the Silent Valley Project, coordinating his international plea with the activities of the domestic Save Silent Valley Committee in India.

This transnationally allied multi-level lobbying began to work, convincing Gandhi to utilize what authority she could wield as a Prime Minister in India's democracy to stall the project. Initially, as State-level elections had yet to be held, the federal Government under Gandhi instructed project authorities to halt work until a new State government was constituted. In the elections held two weeks later, the Communist Party Marxist (CPM) came back to power in Kerala at the head of a coalition of leftist parties. The CPM and its leader, the new Kerala Chief Minister E. K. Nayanar, argued that the Silent Valley Project was critical to the development of the most backward region in the State, which was also one of the CPM's strongholds. The new CPM-led Government of Kerala subsequently attempted to push the project ahead despite Prime Minister Gandhi's opposition. In May, 1980, the Kerala Cabinet under Chief Minister Nayanar publicly announced its intention to move forward with implementation.

The prospects of a politically damaging Centre-State controversy compelled Gandhi to find alternative means to derail the project. Chief Minister Nayanar wrote to Gandhi, stating that he did not want confrontation between his State and the federal government. He thus expected her to persuade the federal Planning Commission to clear the project and also allocate funds for it. In a subsequent meeting between Nayanar and the Prime Minister in Delhi, it was decided that a joint committee consisting of four members nominated by the federal government and four appointed by Kerala State authorities would conduct a third, comprehensive governmental assessment of the Silent Valley Project. Although the committee was to complete its report within three months, it took almost three years.

In addition, the Department of Environment, created from the NCEPC by Gandhi in November, 1980, announced a long-term study to investigate the environmental impacts of three river basin projects in Kerala, including the Silent Valley. The next month, the Kerala Government issued a notification under the federal Wildlife Protection Act that had been passed in 1974 constituting the Silent Valley and surrounding forests as a National Park. However, at that point, the lands to be submerged by the proposed dam project were left out of the area demarcated for the National Park.

Nevertheless, the conflict over the Silent Valley Project had reached a stalemate, which in effect meant that the opposition had succeeded. Project authorities were prevented from moving ahead with construction, and State-level political leaders were frustrated by their inability to fund the project without the support of the federal government. Anti-dam proponents, including the KSSP and others at the local level, the Save Silent Valley Committee, and transnational nongovernmental organizations such as the WWF and IUCN, kept a close watch over the situation. Finally, in the summer of 1983, the joint federal-State committee report on the Silent Valley was completed. It recommended that it was in the public interest to conserve the Silent Valley Forest.[56] In November, after numerous discussions with Prime Minister Gandhi and her representatives, Kerala State authorities announced their decision to stop the project for good and by 1984 the areas to be submerged by the dam were incorporated into the Silent Valley National Park.

The Silent Valley victory inspired big dam opponents throughout India (and all over the world as demonstrated in chapter 6). The momentum gained carried through to a campaign against a series of hydro-projects planned for the Godavari and Indravati rivers in central India. The Bodhghat, Inchampalli, and Bhopalpatnam dams were to displace over 110,000 tribals and flood many tens of thousands of hectares of forests, including part of the Indravati Tiger Preserve sanctuary and the last satisfactory habitat of the Indian wild buffalo. But domestic and foreign environmentalists together with indigenous and human rights activists worked together with the mostly tribal local population to compel the World Bank to suspend its support for the dams, and for the projects to eventually be halted by the Indian State Governments of Madhya Pradesh and Orissa.

The Bodhghat Project, for example, brought together all of these issues and factors. It was to be constructed in the primarily tribal area of Bastar District in the State of Madhya Pradesh as early as 1965 but implementation did not get started for another ten years after that. The first action taken by State officials after restarting the project was to sanction the State Forest Department's proposal to cut down forest composed of sal and teak trees that would eventually be destroyed in the submergence area of the project.[57] No heed was paid to the environmental effects that would be caused by the felling of the trees and clearance of such a vast tract of forest area. Nor were the negative impacts on the approximately 15,000 tribals to be displaced by the dam investigated. Indeed, the Bodhghat was quickly given a clearance at the federal level in 1979 by then Prime Minister Morarji Desai.[58] At that point, the application for an environmental clearance was not mandatory and resettlement policy reforms protecting the rights of displaced peoples had just begun to get incorporated into norms and institutions in India.

But the emergence of grassroots protest, along with the increasing institutionalization and strength of India's environmental bureaucracy and the construction of linkages to transnational nongovernmental organizations, all contributed to the demise of the Bhodghat Project. Babe Amte, a prominent Gandhian activist, who had been initiated into nonviolent campaigning during the struggle for Independence, took up the fight against these big dam projects in the early 1980s, sending numerous letters pressuring Prime Minister Indira

Gandhi to personally intervene to stop construction. In one response, Gandhi replied, "My own views are well known. But it is a very difficult battle. We shall pursue the matter. I am asking the Planning Commission to look very carefully into the aspects you have mentioned."[59]

At the same time, most of the potentially affected tribal peoples joined Baba Amte and commenced a grassroots, nonviolent campaign against the dam, refusing to leave their lands or homes and demanding for the project to be cancelled. As one tribal later remembered, "I am proud of my way of life. Here we are living on our own land. . . . We did not want to be resettled into villages where we all live in rows," while another asserted the degree of dedication to the resistance effort, "We would not go. We were committed to waiting here with our children and our wives until the waters overwhelm us and we will die here."[60]

Simultaneously, domestic opponents of the project began to make use of India's Forest Conservation Act, which had just been enacted by Parliament in 1980 after Indira Gandhi's re-election, to block construction. As a result of the mobilization and lobbying on environmental grounds, a committee within the Department of Environment was constituted to investigate the project with its Secretary, T. N. Seshan, as chair. Based on studies that had been completed, such as one by the Central Forest Institute that found that more than ten million trees would be destroyed, Seshan and his committee withheld the project's environmental clearance for a number of years.[61]

Despite not being sanctioned by India's Department of Environment, the World Bank agreed to support the project by approving loans of $300.4 million. Nearly 60 percent of the total costs were thus covered by the Bank, based on a cursory technical and economic evaluation. The Bank did not investigate either the environmental effects or the displacement consequences before signing agreements with the federal and State governments in India, even though by this time it had also incorporated policy guidelines in these areas (see chapters 3, 4 and 6).

But continued grassroots mobilization assisted by the lobbying of numerous domestic and transnational nongovernmental organizations compelled the World Bank to withdraw its funding offer in 1988.[62] The latter included the Bombay Natural Historical Society in India, as well as the World Wildlife Fund and the prestigious Survival International, based in the United Kingdom, which pressured the World Bank to protect the human rights of the tribal peoples to be displaced by the Bodhghat Dam. The project, effectively suspended as of 1988, was finally cancelled by the State Government of Madhya Pradesh in 1995.[63]

The case of the Subarnarekha Project in the State of Bihar further demonstrates the changing dynamics of—in particular, the growing domestic resistance to and transnationally allied opposition against—big dam building in India from the 1970s on.[64] In March 1975, as soon as preliminary work began on the project, leaders of the tribal people from Singhbhum that were to be displaced protested at the local governmental office. During the ensuing two years, work on the project pushed forward while opposition continued to grow and intensify. In March 1978, an estimated 10,000 people demonstrated against the construction of the Chandil Dam, the first of the two big dams planned for the Subarnarekha Project. The next month, hundreds of villagers began a fast unto the death at the dam site to nonviolently protest against the project. Police attacked the fasters and arrested them. Over 8,000 villagers returned to the site the next day vowing to continue the struggle. The police also returned; they opened fire on the crowd and killed four protesters.[65]

Construction continued on the Chandil Dam, as did repression of the opposition to it. In 1980, the World Bank began preparing a $127 million loan package in support of the project. By the time the Bank was completing its appraisal, the resistance against the Icha Dam, the second big dam in the broader Subarnarekha Project, had come into full swing and was similarly being met by ruthless government repression. The tribal people of the Kolhan area to be submerged by the Icha Project demanded land for land and higher rates of cash compensation. In 1985, a local member of parliament, along with a number of village leaders, sent an appeal to Prime Minister Indira Gandhi to cancel the project.

Despite the continuing protests by the villagers, the World Bank disbursed funds, and by the summer of 1988 the gates of the Chandil Dam had been closed. Ten thousand poor tribals were displaced and another 30,000 were facing the same future as the reservoir filled.[66] Articles in the Indian press and reports written by scholars concurred that the situation had the makings of a humanitarian disaster. After visiting the project site, one World Bank official suggested that the intervention of the International Red Cross might be warranted.

The Word Bank responded to the increasing anti-dam opposition by finally suspending its loan disbursements on October 8, 1988, only to renew its support in 1990, stating that the resettlement situation had sufficiently improved.[67] Domestic nongovernmental organizations, local researchers, and government activists disagreed, arguing that project authorities had still not complied with either domestic Indian or World Bank resettlement policies and that basic human rights standards were being violated. Moreover, despite the existence of legal requirements in India and the World Bank's own guidelines that had been issued in 1984, environmental impact assessments had still not been conducted on the Subarnarekha Projects when the World Bank resumed funding in 1990.

Local opposition to the project did not wane, and the struggle had by then drawn national and international attention and support. On April 5, 1991, 500 people staged a sit-in at the Icha Dam site, protesting against the likely displacement of 30,000 people from 61 villages. The police responded by arresting about one-half of the protesters. It was one month later, on May 13, after more than 10,000 people marched on Patna, capital of Bihar State, that most of those arrested were finally released. A range of domestic groups, led by the Public Interest Research Group and Voluntary Action Network in New Delhi, responded to the events by issuing a statement criticizing not only the police violations of the democratic rights of assembly and organization enshrined in India's Constitution, but also condemning the Subarnarekha Project as a way for officials and contractors to become rich.[68]

The involvement of foreign nongovernmental organizations, linked transnationally to the grassroots mobilization via domestic nongovernmental organizations, was critical to altering the trajectory of the project. In December 1990, a university researcher from the United States visited the submergence area in India and wrote a highly critical report of the project.[69] Based on that analysis, as well as the reports and communications of the domestic nongovernmental organizations in India, Lori Udall of the Washington, D.C.,-based Environmental Defense Fund sent a scathing letter to Heinz Virgin, the Director of the India Department at the World Bank. She wrote about the widespread social and environmental problems associated with the projects, as well as about the repressive tactics undertaken by Indian authorities to quash the opposition. She urged the World Bank to withdraw its support quickly and fully.[70]

The international pressure continued when, at a United States Senate Committee on Appropriations Subcommittee on Foreign Operations hearing, Bruce Rich of the then Environmental Defense Fund conveyed the horrific story of the Subarnarekha Project, and the contributing role of the World Bank, to U.S. Senators who, in turn, questioned the Bank on its handling of the matter.[71] Probe International, a Canadian nongovernmental organization, issued a press release in September 1991, once again decrying the project and the World Bank's support of it.[72] As a result of continuing resistance from local peoples supported by and linked to the lobbying of domestic Indian and foreign nongovernmental organizations, the World Bank withdrew two additional loans it was preparing to assist the Government of Bihar State in completing the project.

Another anti-dam campaign that has been waged against the Tehri Project further demonstrates the changing dynamics around big dam building in India and around the world that has occurred since the early 1970s. Once again, had the usual dysfunctionalities associated with India's policy formulation and implementation processes not delayed the project for so long, it probably would never had been caught in the set of transnationalized conflicts that later embroiled it.

But the anti-Tehri struggle was not as effective due partially to the minimal direct involvement of foreign and transnational nongovernmental organizations in the campaign. Nevertheless, although the project was not permanently halted, the dam was stalled and reformed several times as a result of the sustained, domestic campaign that effectively utilized India's democratic institutions—particularly the courts—and as a result of being strengthened by the domestic institutionalization of novel environmental and human rights norms in India.

Like many similar projects in India, the idea to build a big dam in the northern Indian State of Uttar Pradesh on the Bhagirathi River, which along with the Alaknanda River combines to form the Ganga—perhaps India's most spiritual river—dates back to late colonial/early independence period.[73] The project was not pursued, however, until 1961 when four sites were proposed by the Central Water and Power Commission and a location in the Tehri-Garhwal district was selected. Additional technical studies were conducted over the next eight years and an ostensibly complete project report was submitted to the Uttar Pradesh State Irrigation Department in 1969. The federal Government's Planning Commission sanctioned the project in 1972, based on that report, which consisted of a 260.5 meter dam that would be the world's fifth tallest, if completed. However, the Uttar Pradesh State government did not approve the project for another four years.[74]

By the time authorities were actually ready and able to start implementation on the Tehri Dam in 1976, the political context had changed dramatically. Anti-dam resistance had emerged, in particular, as a result of the grassroots mobilization that was spawned in Uttar Pradesh by the now famous Chipko Movement.[75] The Chipko Movement, which has often been referred to as the people's struggle that initiated India's contemporary environmental movement, began when the Government of Uttar Pradesh State refused to grant control over the forests to local peoples. In October 1971, a large rally against the State Forest Department's decision to auction the forests to outside contractors was coordinated by a grassroots group that had been formed to assist local communities. This demonstration was followed by public meetings among the local peoples that resulted in demands for traditional forest rights and support for organized forest labor cooperatives. Subsequently, the Forest Department did not grant permission to the local peoples to plant ash trees, while allowing the same type of trees to be felled in a nearby location by a sports company from outside the area.

The Chipko Movement got its name because the local villagers chose to protest the hypocrisy of the State government and State Forest Department by preventing the sports company or any other outside contractor from felling any trees in the area. They did this by staging the tremendously symbolic act of

Chipko—hugging the trees to pose a human barrier between the forest to be cut down and the lumbermen. The State Forest Department was undeterred by a series of such demonstrations, however, and continued to auction the trees off to the highest commercial bidder. In 1974, after tricking the men-folk out of the villages, the Forest Department sent lumbermen into the forests at Reni to fell trees. But the women of Reni responded quickly, mobilizing themselves to non-violently resist the lumbermen by hugging the trees—Chipko. The women stood their ground, despite being threatened, and the lumbermen were forced to withdraw.[76] As a result of the Reni episode, the Government of Uttar Pradesh agreed to constitute a committee to investigate the situation. The committee's report submitted in 1977 legitimized claims that had been lodged by the Chipko Movement about the destructive consequences of commercial deforestation on the fragile Himalayan ecosystem.

The Chipko Movement then spread to the Tehri-Garwhal area and provided the base upon which the anti-Tehri dam campaign was initiated. Sunderlal Bahaguna, a well-known Indian activist, had been involved with the Chipko struggle from its outset. At the same time, he became gravely concerned about the proposed construction of the Tehri Dam. The various criticisms of the project that were made by Bahugana and others—about dam safety given that the site was located in an earthquake prone area, lack of rehabilitation of peoples to be displaced, and loss of even more forests in this primarily mountainous area—galvanized further grassroots mobilization against the dam among the local peoples who were already organized in the Chipko movement.

As a result, the struggle against the Tehri Dam grew in strength. By 1978, numerous meetings of local village councils (gaon sabhas) and the district council (zilla parishad) led to the formation of the Anti-Tehri Dam Struggle Association (Tehri Bandh Virodhi Sangharsh Samiti). Protests in the form of rallies, demonstrations, and fasts were coordinated to prevent construction of the project, and a petition was presented to India's federal parliament in 1979. In response to the growing criticism and to further justify the construction of the Tehri Dam, the Uttar Pradesh Irrigation Department submitted a revised project report that same year, greatly expanding the scheme. The revised Tehri Dam Complex was now to generate 2400 megawatts of power, as opposed to originally proposed 600 megawatts, as well as supply additional irrigation potential to increase the benefits produced relative to the costs.

But, as in the case of the Silent Valley Project, the creation of the Department of Environment in 1980 shifted the dynamics around the Tehri Dam. Pressured to respond to the anti-dam mobilization, and the personal letters sent by Sunderlal Bahugana to Indira Gandhi, asking her to halt the project, the federal government constituted a working group under the auspices of the Department of

Environment to evaluate the environmental effects of the Tehri Dam shortly after Gandhi returned to power. As Prime Minister Gandhi wrote at the time in a letter to the Department of Science and Technology, "There are several proposals (for big dams) which were agreed to earlier but would need to be looked into again. Among them are Silent Valley, the dam in Tehri Gurhwal and the dam in Lalpur, Gujarat. . . . there is great local distress and a feeling that contractors and other such groups will be the main gainers. Hence it is necessary to have another look in depth."[77]

The growing bureaucratic conflicts among the various federal agencies over the project were readily apparent when an interim report of this first federal government working group on the Tehri Dam was submitted in May 1980. Members from the new Department of Environment suggested that no recommendation could be made without further investigations to fill in the wide gaps in data. However, the committee members representing the Department of Power and the Central Water Commission argued that the project was well formulated and that construction should not be delayed.

These inter-bureaucratic conflicts, generated by the creation of the Department of Environment, would contribute to altering the trajectory of the Tehri Project throughout the 1980s and 1990s. For example, the studies to be conducted by the Department of Power, Central Water Commission, and Uttar Pradesh Irrigation Department, asked for by the environmental members of the committee, were still not completed six years after they had been requested. Consequently, in August of 1986, the chairman rejected the draft report prepared by the working group because of its weak empirical foundation and unilaterally submitted an independent report on the project. In the cover letter addressed to T. N. Seshan, the then Secretary of the Department of Environment and Forests, Chairman S. K. Roy, wrote:

> I regret having to record my considered view that virtually no importance has been given to the many varied and significant aspects of the complex relationship between the construction of a high dam on one of the world's most important rivers, and either the human or natural environmental impacts . . . there was a deep difference of opinion within the Working Group. . . . Inevitably, it forced a toning down of the environmentally based, less technologically measurable factors and led to the compromise . . . that the Working Group would not make any definite recommendation on stoppage of work on the Tehri Dam. . . . I confess that my experience in this case as Chairman has been harrowing and distressing. I have chaired innumerable committees and groups in India, and in other parts of the world. I have never encountered such an unbending dogmatic approach to all issues which were not positively framed to ensure continued work on the Tehri Dam, whatever the cost to the environment and public funds. . . .[78]

With the report, as a proponent of the Tehri Dam later remarked: "The breach was complete. The Ministry of Environment and Forests on the one hand and the Department of Power with the Central Water Commission on the other appeared equally divided [on the viability of the project]."[79]

The domestic struggle against the Tehri Project mounted during the period in which the working group was mired in these bureaucratic conflicts and the judicial process within India's democratic system became a primary locus of the anti-dam campaign. Becoming more anxious by the delay in submission of the working group's report, anti-Tehri actors submitted various cases against the project to India's judicial system, from the local courts to the High Court of Uttar Pradesh State and ultimately to the Supreme Court in New Delhi.[80] A writ petition, filed before the Supreme Court in 1985, argued that: "The Tehri Dam Project is a very dangerous venture, it is technically infeasible, geologically a blunder, economically unsound and environmentally disastrous. It is not a project of development but a scheme for destruction. . . ." More specifically, the petition stated that, "The chances of failure of the dam (from an earthquake) are too great and real and, if that evil occurs, all habitations . . . will be wiped out of existence, causing a disaster unprecedented in the history of man. It is claimed that the Respondents have no right in law or otherwise to make such massive inroads into the fragile eco-system of the Garhwal Himalaya. . . ."[81] The Indian National Trust for Art and Culture (INTACH), a nongovernmental organization located in Delhi, filed an application in support of the 1985 writ petition claiming that the Tehri Project violated the Right to Life of the people enshrined in Article 21 of the Indian Constitution.[82]

But the combined environmental and human rights arguments of the anti-dam proponents did not convince the Supreme Court. Hearings were held on the case over a five-year period, during which federal and State authorities countered the arguments of the petitioners. The judges ruled in November 1990 to dismiss the claim, because the Court did "not possess the requisite expertise to render a final opinion on the rival contentions of experts" with respect to technical and environmental issues. The justices noted further that they were satisfied that the federal and State authorities had moved ahead with the Tehri Project only after making an informed decision on the matter, and thus these authorities had not violated the fundamental rights of the people affected by the project.[83]

While the Supreme Court hearings were under way, domestic authorities searched abroad for financial and technical project support. In 1986, an Indo-Soviet accord was reached, through which the Soviets offered one billion rubles in financial assistance, as well as implementation of the power component and some other technical aspects of the project. After conducting numerous investigations the Soviets stated, in an attempt to contribute to positive opinion about the project within India, that the dam design was safe.[84]

The collapse of the Soviet Union in 1989, however, derailed the Soviet financial and technical support Indian authorities were counting on. The subsequent political and economic problems faced by Russia during the 1990s made it difficult for that Soviet successor state to fulfill the commitments made to India by the former Soviet Union. But domestic and international opposition to the project also contributed to the Russian pullout. Prior to President Boris Yeltsin's visit to India in 1992, INTACH and 54 nongovernmental organizations from India and around the world sent a letter to him urging him not to revive Russian backing for the project. Yeltsin complied, although he had seriously considered the prospect.[85]

Consequently, by 1990, Indian officials and authorities were still faced with a funding shortfall while continuing to face sustained pressure on the potential social problems and environmental dangers of the Tehri Project by anti-dam proponents. After the Environmental Appraisal Committee on River Valley Projects within the Ministry of Environment and Forests refused to grant environmental clearance—as a result of safety issues with respect to the dam in 1989—a High Level Committee of Experts was constituted to review the potential geological and seismic risks of the project. That committee submitted a report in April 1990 stating that the dam was earthquake safe. The Ministry of Environment and Forests was thus compelled to accord a conditional clearance to the Tehri Hydro Development Corporation in July of that same year.[86] The conditions included the formulation of comprehensive environmental management plans, the investigation of rehabilitation measures to ensure that a reasonable quality of life of displaced people would be assured, and the preparation of a disaster action plan in case a dam failure occurred.

Given the environmental clearance and the positive ruling of the Supreme Court in November 1990, project authorities believed that opposition to the Tehri Dam would subside. However, the combination of pressure by Delhi-based nongovernmental organizations and grassroots resistance continued to prevent completion of the project. INTACH lobbied the Environmental Ministry to halt work on the Tehri Dam in September 1991, since the requirements laid down in the conditional environmental clearance had not been fulfilled. Then, the following month, the massive 6.1-Richter-magnitude Uttarkashi earthquake rocked areas nearby the Tehri site. This sparked a re-intensification of the dam opponents' lobbying and mobilization efforts. By December, two Delhi-based environmental critics submitted another writ petition in India's Supreme Court against the project, again based on the potentially affected peoples' constitutional right to life and right to know.[87]

The local people also responded immediately after the earthquake with a strong grassroots mobilization effort of their own, marching to and taking over the dam site, and halting construction work. After 75 days of stalemate, the protesters were arrested. Their leader, Sunderlal Bahuguna, subsequently initiated an indefinite fast demanding an independent review of the project. On the 45th day of the fast, Prime Minister Narishima Rao agreed to have the project reviewed but the anti-Tehri actors still refused to call off the campaign. In Delhi, concerned individuals and nongovernmental organizations formed the Tehri Action Group to sustain the pressure on federal government and bureaucratic authorities and to ensure that work would not restart without a report prepared by an independent review team.[88]

But the federal government did not keep its promise, despite the second Supreme Court case it faced, the lobbying activities of the Tehri Action Group, and the persistent resistance by local people. Bahuguna subsequently conducted two more fasts, one of 49 days in 1995 and 75 days in 1996.[89] The federal government responded by constituting two more committees to review the safety and rehabilitation aspects of the project, once again delaying work indefinitely.[90]

Thus, after twenty-five years of resistance, Indian authorities still did not announce the cancellation of the Tehri Project. Yet, the dam had not been completed either, while substantial reforms had been adopted. The absence of foreign financial and technical support minimized the role that lobbying efforts by foreign and transnational nongovernmental organizations could play in this struggle.[91] Nevertheless, transnationally allied opponents of the project did organize letter-writing campaigns to project authorities criticizing the Tehri Project and have provided important technical and scientific information to support domestic anti-dam actors.[92] Moreover, globalizing environmental and human rights norms that were domestically institutionalized played a central role in providing opportunities to the dam opposition within India's democratic system. Ultimately, like most resistance against big dam building in India, the effects of the campaign against the Tehri Project depended upon the sustained and organized grassroots mobilization and the existence of domestic democratic institutions.

DAMS, DEMOCRACY, AND DEVELOPMENT IN INDIA

The growth and intensification of a domestic debate over big dams in India were linked to the emergence of various anti-dam struggles since the early 1970s. Within a decade, public criticism of the unfulfilled benefits and spiraling costs of big dams in India had mounted and contributed to both the formation of a broader anti-dam movement domestically as well as the transnationally allied opposition to big dams around the world. Perhaps the first of these critiques to gain widespread domestic attention was *Major Dams: A Second Look,* edited by L. T. Sharma and Ravi Sharma. The volume was published in 1981 by the Gandhi Peace Foundation, a Delhi-based civil and human rights nongovernmental organization that had just formed an environmental unit and focused on the negative social and environmental consequences of big dam building in India.[93]

The Centre for Science and Environment, another prestigious Delhi-based nongovernmental organization established in 1980 and subsequently developing extensive transnational linkages, published its first *State of India's Environment-1982: A Citizens' Report,* a year later. The report systematically overviewed environmental problems in India, including those related to big dams. The debate over these projects grew, working its way into the Indian national press with such articles as Ravi Sharma's "Real Cost of Big Dams" and "Need for Frank Discussion," as well as with Darryl D'Monte's "High Dams at a High Price."[94] By the mid-1980s, there was widespread domestic media coverage on the subject.

Criticism of big dam building in India was also being leveled outside the country. For example, a report entitled "Dams in India" was written by the British nongovernmental development organization, Oxfam, in 1983. The next year, Edward Goldsmith and Nicholas Hildyard highlighted the negative consequences of the Indian big-dam building experience in the first volume of their internationally publicized *The Social and Environmental Effects of Large Dams.* An entire section in volume 2 was devoted to case studies of big dam projects in India, including articles on the Tehri and Narmada Projects written primarily by Indian critics.[95] Subsequently, in September 1987, the Society for Participatory Research in Asia (PRIA), another New Delhi-based nongovernmental organization, held an international workshop on big dams in India. The workshop, which was attended by scholars and activists from all over India and across Asia including Malaysia, the Philippines, Sri Lanka, and Thailand, led to PRIA's publication of *Peoples and Dams* in the late 1980s.[96] Finally, critics of India's big dam building practices lobbied foreign governments and international development agencies for the withdrawal of financial and technical support for these projects during the 1980s, in conjunction with the formation of various transnational anti-dam campaigns.[97]

As a result of the growing public debate and campaigning around the world, mounting struggles against these projects domestically, and emergence of transnationally allied opposition, big dam building became one of the most politicized development activities in India by the 1980s. By 1986, even Prime Minster

Rajiv Gandhi reflected negatively on India's historical record with major irrigation projects, generally involving the construction of big dams: "We can safely say that no benefit has come to the people . . . We have poured money out, the people have nothing back: no irrigation, no increase in production, no help in their daily life."[98] Three years later, former federal Secretary of Water Resources Ramaswamy Iyer wrote in a widely read article[99]:

> We should certainly accord priority to the utilisation of the potential already created, the reclamation of the potential which has been lost through misuse, and a vast improvement in water management. . . . We should also place a much greater emphasis than in the past on minor irrigation, which calls for less immediate investment, promises quicker results, and presents fewer problems . . . considering the heavy costs (financial, human, social and environmental) involved in large-dam projects, we have to be highly selective and extremely cautious regarding approvals to such projects.

This was a dramatic shift for Iyer, who in 1987, played a central role in convincing Prime Minister Rajiv Gandhi to grant an environmental clearance so that construction could begin on the controversial Sardar Sarovar major dam component of the larger Narmada Projects (see chapter 4).

A collection of essays written by both proponents and opponents crystallizing the big dams debate in India was incorporated into B. D. Dhawan's edited volume, *Big Dams: Claims, Counter-Claims* in 1990. While Dhawan "hoped that this would generate wider and more informed participation in the controversy, and hopefully to its resolution," the papers more strongly demonstrated the growing gulf between proponents and opponents of these projects.[100] In fact, B. G. Verghese, a former member of India's Planning Commission, vociferously defended big dams in a book published in 1994 concluding that, "large water projects that have come under increasing attack are not monumental follies. They offer hope to millions. India has to be bold and more active, willing and able to reach out and grasp opportunities and not shrink from them . . . Narmada, Tehri, and the IGNP, like Bhakra-Pong are milestones along the way."[101]

Nevertheless, by that time, virtually every such project in India, whether at the planning stage, being implemented, or even completed, was being publicly criticized and/or actively opposed. As a result, the tide had visibly turned to the side of the opponents against big dams in India. As one former Chairman of the federal Government's Central Water Commission expressed in dismay, "If the Bhakra-Nangal Project was proposed today, it would never be completed. There is just too much opposition."[102] India's Power Minister, N. K. P. Salve, put it bluntly when he stated in 1993, "We are not going in for large dams any more. We want run-of-the-river projects and to have smaller dams, if they are necessary at all, which will not cause any impediment whatsoever to environmental needs."[103] This profound change was almost a complete reversal from the early years of India's Independence when Nehru idealistically and confidently called big dams the "modern temples of India."

The accumulation of anti-dam struggles during this period clearly spurred the formation of a broader anti-dam movement in India. In 1988, for example, a meeting of over 80 prominent scholars, activists, critics, and leaders of struggles, including Sunderlal Bahuguna from the anti-Tehri campaign and Medha Patkar from the anti-Narmada campaign, was organized by the renowned Gandhian Baba Amte, who himself had lead the anti-Subarnarekha campaign.[104] As a result of their discussions, the participants proclaimed the following "Appeal to the Nation":

> Around eighty of us met . . . to express our grave concern about the devastation caused by big dams. We came from different parts of the country, all united by a common resolve—to ensure that people were no longer denied their basic rights over natural resources. We affirmed that the nation's rivers are the cradle of our civilisation and that they cannot be strangulated to meet the needs of the exploiting class within society. The issues raised by the construction of big dams challenge the very concept of the present pattern of growth, unquestioningly adopted by our planners. Nothing less than the survival of life itself is at stake for very many of our people, and time is running out rapidly. We appeal to the Nation to halt all big dams, here and now.[105]

Moreover, in an "Assertion of Collective Will Against Big Dams," the group not only called for a moratorium on big dam building but proposed that "ultimately, the entire planning process has to be reversed, whereby each village becomes a unit, and decisions regarding its development are made by the people, taking the resources provided by the specific ecosystem into account. . . . Development and the protection of environment will then proceed hand in hand, and industrialization will grow in response to people's demands."[106]

Anti-dam struggles and the broader anti-dam movement have thus not only campaigned against the construction of big dam projects, they also promoted different activities, policies, strategies, and an entirely different vision of development for India than the one that had been followed since Independence. Besides reforming, stalling, and even halting big dam building, anti-dam proponents were a major factor in development policy reforms in areas including water and energy, forests and resettlement.[107] For example, the prioritizing of the provision of drinking water and the creation of a National Drinking Water Mission in 1986, as well as the formulation of a National Water Policy in 1987, and subsequent debates over it, were clearly influenced by criticisms of big dam building.[108] Similarly, since 1976 when the Government of Maharashtra State was compelled to enact a "Displaced Persons Rehabilitation Act" in response to a number of agitations that occurred during the 1970s against big dams that had been built in that State, big dam critics have lobbied intensively for policy enactments and changes in this area.[109] Years of campaigning resulted in the 1994 drafting of a "National Policy for the Rehabilitation of Persons" by the federal Ministry of Rural Development that, while not deemed acceptable in that form by many at the time, provides further evidence of the increasing power and effectiveness of anti-dam struggles.[110]

Finally, the anti-dam movement has been one of the leading forces and constituent elements in the growth of a broader sustainable human development movement in India. On 18 September 1989, the first "National Rally Against Destructive Development" was held in the town of Harsud, a village to be displaced by one of the big dams proposed for the Narmada River Valley in the State of Madhya Pradesh. The between 50,000 and 100,000 people who attended the rally included not only tribals, peasants, activists, and leaders fighting against big dams, but thousands of others who believed that India had to fundamentally alter its model of development.[111] Following the Harsud Rally, environmental and social activists from all over India met in Bhopal, Madhya Pradesh and founded the Jan Vikas Andolan, or Movement Against Destructive Development.[112] Building on the Jan Vikas Andolan, anti-dam actors linked up with environmental, tribal, lower caste, women's, human rights, and other like-minded groups such as the National Fisherworker's Forum to establish in 1993 the National Association of People's Movements (NAPM).[113] The NAPM was dedicated to building a powerful, nonparty, political force around which the thousands of domestic nongovernmental organizations that proliferated since the 1970s could unite with historically marginalized social groups and people's movements to promote an alternative vision of development for India.[114]

However, three trends that became more prominent during the 1990s did begin to counter the strength and undermine the effectiveness of big dam opponents, sustainable human development proponents, and their transnational allies in India. The first was the progressively more active organization and mobilization of pro-dam actors, an indication in itself of the changed dynamics of development in India. These have included such groups as rich farmers and middle classes; technical, professional, and bureaucratic elites linked to big dam building; private sector companies who increasingly wanted to seize new profit-making opportunities linked to these projects; as well as politicians who were dependent on these groups for their political survival and success. These actors also strengthened their transnational linkages to like-minded groups from international professional associations, such as the International Commission on Irrigation and Drainage; allies in donor agencies, like the World Bank; multinational corporations; and nonresident Indians living abroad.

The second trend was the progressive adoption of neo-liberal economic policies and the related privatization and liberalization of the power and water sectors similar to most developing countries during this period.[115] The policy and institutional changes that had been promoted, and to a certain extent achieved, with respect to big dam building by opponents had not focused on the increasing private sector involvement in big dam building and operation around the world and in India. Big dam critics thus increasingly faced less conducive political opportunity structures than if public authorities were the lead developers of these projects. But as the conflicts over the Maheshwar Dam component of the Narmada Projects that domestic Indian private sector firm S. Kumars took over building demonstrated, if foreign corporations are actively involved, then transnationally allied opposition linked to domestic mobilization is likely to increase and make the building of a big dam extremely difficult in India.[116]

The third trend was the growing strength of Hindu fundamentalist religious movements and right wing political parties within India. Extremist organizations, such as the Rashtriya Swayamsevak Sangh (RSS), Vishwa Hindu Parishad (VHP), and others that are collectively known as the Sangh Parivar increasingly and successfully "Hinduised" marginalized groups such as tribal adivasis and lower castes by offering services, on the one hand, and utilizing force and terror tactics, on the other. They also contributed to the election of the right wing Bharatiya Janata Party (BJP) in numerous states and at the federal level in India by the second half of the 1990s. The consequences included progressively more violent and repressive politics in both society and the state and a weakening of the country's democratic regime. And as democratic political opportunity structures were undermined, the likelihood that the changing dynamics of big dam building and development (as identified in this chapter) could be sustained decreased.[117]

Nevertheless, as has been demonstrated here, from the 1970s to the 1990s, previously marginalized, domestic subaltern groups, such as poor peasants and tribals, have become progressively more aware, empowered, and organized in order to mobilize for the substantial reform, if not cancellation, of big dams, and to lobby for greater access to India's democracy and greater inclusion in India's development. The mobilization by these subaltern groups was ardently and consistently supported by the tremendous proliferation of activists and nongovernmental organizations committed to promoting transparency, participation, cultural diversity, social justice, and environmental sustainability. The resultant coalitions and networks were transnationalized through links with like-minded foreign allies and through legitimation by the growing domestic institutionalizing of globally spreading norms. The in-depth historical and ethnographic analysis of India's Narmada Projects offered in chapters 3 and 4 along with the cases of other developing countries analyzed in chapter 5 will provide further evidence that these combined factors often result in the reform, stalling and/or halting of big dam building. They also have been shaped by and contributed to the transformation in the transnational political economy of development in India, around the world, and at the international level.

❖

CHAPTER THREE

India's Narmada Projects

Historical Genesis and the First Transnational Reform Campaign

The fact that Indian authorities planned numerous big dams for the Narmada River Valley as far back as 1946 and that, with a few exceptions, few of these or other projects planned subsequently were completed by the early 1990s, forms the basis of the comparative historical and ethnographic examination of India's Narmada Projects presented in the next two chapters.[1] Indeed over 3,000 dam projects, including 165 big dams, were finally approved for construction along the Narmada River's course and tributaries in 1979, constituting part of the largest river basin scheme formulated in India since Independence.

The analysis developed in this chapter shows that transnational opposition, norms for environmental protection, and the heightened focus on human and indigenous peoples' rights did not exist either globally or in India and therefore did not prevent construction of the big dams in the Narmada River Valley that were proposed during the first decades after Independence. I also demonstrate that domestic social mobilization against the proposed big dams was not generated during this initial period and, when it finally did emerge in the late 1970s and early 1980s, was largely ineffective in independently altering the course of projects proposed for the Narmada River Valley. Rather, the initial delay of more than three decades in sanctioning the Narmada Projects was primarily the result of political and institutional dynamics within India's federal bureaucratic democracy, similar to those identified in the previous chapter.

It was only after the emergence of a first transnational coalition and a five-year, multilevel advocacy campaign was conducted to defend the rights of people to not be negatively affected by the Sardar Sarovar major dam component of the scheme that pro-dam actors such as Indian and World Bank authorities visibly reformed their social policies in the areas of indigenous/tribal peoples and resettlement with respect to the project. But environmental criticisms were not integrated into this primarily resettlement reform campaign to the extent that they were in the second transnational coalition that subsequently emerged to completely oppose the Narmada Projects—the focus of chapter 4.

The next three sections demonstrate that institutional bottlenecks and political conflicts specific to post-colonial India's evolving federal bureaucratic democratic system, as well the quest by Independent Indian authorities to build more and ever larger projects that would ostensibly generate greater benefits given the hegemonic development vision were the primary causes of a more than thirty-year delay in even initiating most of the Narmada Projects. I first detail the origins of planning for the Narmada River Valley from the initial investigations prior to Independence to the various plans and projects proposed for the basin between 1946 and 1960. I then trace the emergence of conflicts between the riparian regional States, and related formulation of ever more ambitious development plans culminating with the scheme approved for the basin in 1979, that basically stalled big dam building for two decades.

This historical overview of the decision-making process and outcomes during these initial two periods reveals factors that conditioned the increasingly transnationalized struggles over the Narmada Projects during the 1980s and 1990s. The next section thus offers an account of why independent domestic resistance that emerged right around the time the 1979 scheme for the basin was approved ultimately proved rather ineffective. This relative failure is then contrasted in the last parts of the chapter with the subsequently discernible effects produced by the first transnationally allied campaign to reform the resettlement policies and practices around the Sardar Sarovar major dam component of the Narmada Projects.

EARLY ATTEMPTS TO DEVELOP THE NARMADA RIVER VALLEY

As with other rivers in India, references to irrigation development abound in the folklore and ancient writings about the Narmada.[2] With the advent of British rule, however, irrigation in India attained an even greater prominence. Although much was done to repair and improve some of the most important existing irrigation works that had been indigenously built in the pre-colonial period, The First Famine Commission of 1880 further emphasized the need for direct involvement by colonial authorities in the development of new irrigation projects on the Indian peninsula.

Yet, until the 1890s, the colonial government did not formulate comprehensive plans for any of the rivers on the subcontinent, partly as a result of the good agricultural harvests recorded from 1880 to 1895.[3] The two great famines of 1887–88 and 1899–1900 subsequently compelled the British colonial author-ities to appoint the First Irrigation Commission to report on irrigation as a means for protecting against such disasters in India. The Commission con-ducted a thorough review of irrigation development and examined various pro-posals for new schemes to be built on India's rivers.

But even after the turn of the century and through the end of the colonial pe-riod, no major irrigation works such as canals or big dams were recommended for the Narmada River Valley, even though the British had built more than half of the world's big dams at the time. The Commission members argued that the deep black soil found in the basin would not likely stand up to the continuous irrigation made available by these types of large infrastructure projects. Thus, although a considerable amount of resources went into the promotion of smaller and private irrigation works, little activity in terms of large scale, state-led development in the Narmada River Valley occurred until after World War II.

Under Prime Minister Nehru and the Congress Party's direction, the post-colonial Indian water and power bureaucracy forged ahead with the planning of big dams throughout India, as explained in chapter 2. The first proposal for the integrated development of the Narmada River Valley was made via the In-dian bureaucracy in 1945–46 by Dr. A. N. Khosla, the Chairman of the federal Government of India's Central Waterways, Irrigation and Navigation Commis-sion (CWINC).[4] Correspondingly, the first formal investigations for compre-hensive planning of the basin in the areas of flood control, irrigation, power, and extension of navigation were initiated on the eve of Independence in 1946–47 by the CWINC at the request of the Governments of the Central Provinces and Berar and Bombay Province.[5] The study of the topography and hydrology of the Narmada basin revealed several potentially exploitable sites for big dams. These sites were inspected in a preliminary fashion by engineers and geologists, and further detailed investigations for seven projects were recommended.

The following year, despite uncertainty about the scarcity of human and fi-nancial capital in India, Dr. A. N. Khosla, Dr. J. L. Savage, and M. Narasimhaiya were appointed to an ad hoc committee by the Central Ministry of Works, Mines & Power (CMWMP) to review the preliminary estimates and formulate priorities for the in-depth investigations on development of the Narmada River Valley. The committee recommended that detailed studies be conducted for four of the seven sites that had been proposed—the Bargi, Tawa, and Punasa Projects (the predecessor of the Narmada Sagar major dam Project) in Madhya Pradesh, and Broach Barrage and Canal Project (the predecessor of the Sardar Sarovar major dam Project)—in what was then Bombay State:

The Narmada basin appears to hold great potential for development, and such de-velopment is likely to have far reaching effect on the economic advancement of the country in general, and the basin in particular. . . . We entirely agree with the view that, given shortages of men and materials, it will be desirable to restrict the work in the first 2–3 years to only such projects as will give maximum results in the shortest possible time. Judged from this criterion, we recommend that investigations be concentrated on four projects. . . .[6]

The federal Government of India sanctioned the funds for these investigations and the CMWMP delegated the task to the CWINC in 1949.[7] Reports for the Tawa, Punasa, and Broach Projects were completed over the next half-decade, but the Bargi investigations were suspended due to a lack of funds.

Power, crucial to industrialization and a symbol of modernity from the perspective of Indian authorities, was made a priority alongside irrigation in the development vision and strategy adopted by the Indian state during Nehru's first term in office.[8] The CWINC was thus renamed the Central Waterways & Power Commission (CWPC) in 1951 as part of a bureaucratic and governmental reorganization implemented during the initial years of India's Independence. Correspondingly by 1955, while the other investigations continued, the CWPC completed another study on the hydroelectric potential of the Narmada River and its tributaries, finding that with adequate regulation, approximately 1300 MW of power could be generated across 16 sites. A report was also prepared by the CWPC for the construction of a weir at Gora with pond level of 160 feet for irrigation purposes as part of investigations for the Broach Project in 1956. In February 1957, the Gora site was inspected further by a senior designs and research member of the CWPC. Finding that the studies of the Gora site were not complete, he proposed further analysis at that location as well as at a site 1.5 miles upstream called Navagam.[9]

Given India's adopted development vision and strategy, larger projects were privileged as greater and greater amounts of irrigation and power were always desired. Navagam thus came to be favored by the CWPC, because the exposed rock in the river bed afforded high abutments (lateral rock supports jutting out of the main rock formation) which facilitated raising of the dam height and increasing the potential benefits derived from the Broach Project. Not surprisingly, the Government of Bombay State quickly agreed that Navagam would be the ideal site for a big dam after consultations with the CWPC about the potential for increasing the benefits from the project.

The effect of this project on the State of Madhya Pradesh and broader attempts to develop the basin were not assessed at the time, even though coordinated planning for the Narmada River Valley had been prioritized by the CWPC. On September 24, 1957, representatives of the States of Madhya Pradesh and Bombay—the latter to which the States of Saurashtra and Kachchh had been merged in 1956—and the Chairman of the CWPC did, however, meet once again to discuss the ongoing planning for the comprehensive development of the Narmada River Valley. Based on the accumulation of these studies and reports, it was agreed that three additional intermediate sites between Punasa and Broach—Barwah, Harinphal, and Keli—would be investigated at the shared cost of the two State governments.

But no discussion of the conflicts between the States that potentially would arise from the building of the major Broach dam at Navagam was conducted. Moreover, neither the possible negative social and environmental consequences nor the need for participation by the people to be most affected by the project was raised. The CWPC thus continued its investigations unimpeded and forwarded its report for the revised Broach Irrigation Project at Navagam to the Government of Bombay State in January 1959. Implementation was to proceed in two stages. In the first stage, the dam was to be built to a Full Reservoir Level (FRL) of 160 feet high with provision for wider foundations to enable raising the height to FRL 300 feet in the next stage. Constructing a high level canal in addition to increasing the dam height was proposed for the second stage.

Initiation of the construction of the Broach Project at Navagam was delayed once again, initially as a result of the quest of Indian authorities to increase the size and maximize the generation of benefits from all potential development projects in the Narmada River Valley. On January 16, 1959, the Government of Bombay State conveyed to the CWPC that because the big dam at Navagam was to eventually be constructed to a height of FRL 300, there would be no need for building a big dam at Keli in between Harinphal and Navagam.[10] The Government of Bombay State also proposed raising the ultimate height of the dam even higher, to 320 feet in the second stage. This increased height was to allow for the construction of powerhouses at the head of the low level canal and in the riverbed. This revised, multipurpose Broach Project was reviewed and quickly accepted by the CWPC.[11]

The project was also delayed because of the involvement of multiple authorities and agencies within India's federal bureaucratic democracy in its formulation and implementation. For example, the federal Government of India's Ministry of Irrigation & Power appointed a consultant team to conduct a review of the revised version of the Broach Project. Although the consultants did not submit their report until April of 1960, they eventually agreed that the Navagam Dam in the Broach Project should be constructed to a height of FRL 320 feet in only one stage, rather than two. But they also suggested that irrigation be extended to the drought prone Saurashtra and Kachchh regions of what was then Bombay State, as well. As we shall see, inclusion of these areas would later become a critical factor in the justification and controversies surrounding the

Sardar Sarovar Project, which replaced the Broach Project, during the 1980s and 1990s.

Progress on the Broach Project and Navagam dam was nevertheless set back again due to institutional changes within India's federal bureaucratic and democratic polity. On May 1, 1960, the erstwhile State of Bombay was divided into the States of Maharashtra and Gujarat, even as the suspended investigations of the Bargi Project were re-initiated by the CWPC.[12] As a result of the territorial reorganization, the Navagam site was now located in the State of Gujarat and the planning and implementation of the Broach Project was transferred to this completely new State's first government and practically nonexistent bureaucracy. Nevertheless, the federal Government of India's Planning Commission—the bureaucratic agency with final authority to approve development projects at the time—sanctioned Stage I of a revised Broach Project in the summer of 1960. In fact, the Planning Commission notified the Government of Gujarat's Planning and Development Ministry of its approval in a letter dated August 5, 1960, only a little more than three months after the State of Gujarat itself was constituted.

The Broach Project was sanctioned without it being systematically integrated into a river basin plan and without the agreement of the adjacent riparian State of Madhya Pradesh. Rather, according to the Planning Commission, the Navagam Dam was to be built to a height of FRL 162 feet with a low level canal taking off from it for irrigation purposes in the first stage of the Broach Project. In addition, the plans for the initial stage provided for construction of an ungated weir with FRL 162 feet for diverting water into a low level canal. Only the anticipated releases from the Tawa Project in Madhya Pradesh were partially incorporated into the planning, and the irrigation to be generated was earmarked for the Broach and Baroda districts of Gujarat as primary beneficiaries.[13]

Moreover, neither an economic analysis nor a comprehensive financial plan was formulated for the Broach Project at the time of its sanctioning. The project estimates and costs, as well as planning and layout of the original dam, however, also included obligatory works required for raising the dam height to FRL 320 feet in the second stage. While Stage I of the project focused solely on generating irrigation, over 1,000 MW of power was to be produced from the construction of a riverbed powerhouse in Stage II. Moreover, irrigation was to be extended in the latter phase to the Saurashtra and Kachchh regions of Gujarat by means of a high level canal taking off at a full supply level (FSL) of 295 feet.

Again, examination of the potential negative social and environmental effects of the project had still not been conducted when the Broach Irrigation and Power Project at Navagam was inaugurated by Prime Minister Jawaharlal Nehru in April of 1961, only two months after the Government of new Gujarat State had given its administrative approval. At that time, the first six villages (four completely, two partially) were commandeered for Kevadia Colony, the headquarters for construction of the project. The impacts on those displaced, who were mostly tribal peoples, were not examined. The people who were affected by the project were not informed in advance, nor was their participation considered necessary for successful execution of the projects. A policy of resettlement and rehabilitation was not formulated, nor was compensation awarded for the loss of lands and livelihoods. The "oustees," as they were called, themselves were not organized, mobilized, or even aware of their rights enshrined in India's democratic constitution.

CONTINUING DELAYS IN FORMULATION AND IMPLEMENTATION

When Nehru inaugurated the Broach Project, authorities optimistically expected that it would be completed and commence generating benefits within a period of seven to ten years, that is, before 1970. But, although preliminary works on the project were already under way, the drive to increase the size of projects and their development impact as well as continuing bottlenecks and conflicts internal to India's federal bureaucratic democracy further delayed the formulation and implementation of most projects for the Narmada River Valley.

In particular, increasing inter-State tensions over development of the basin and the failure of federal authorities to mediate these conflicts resulted in the constitution of an inter-state water dispute tribunal to develop an integrated plan for the basin. The tensions were sparked when the Government of the newly created Gujarat State (GOG) initiated studies on alternatives for utilizing the flow of the Narmada River below the Punasa site in Madhya Pradesh. The focus was not on alternatives to big dams, but on even larger and more grandiose projects that would ostensibly more fully exploit the irrigation and hydropower potential of the Narmada Valley to advance economic development. As the Broach Project would be the terminal reservoir along the river, the GOG argued that it should be allowed to build the Navagam dam to store all the water flow in the river available at that point. The survey work relating to the likely submergence area of the reservoir and the area of the high level command from the construction of such a dam was entrusted to the Geological Survey of India in late 1960. The surveys for the construction of a high level canal for increased irrigation delivery were undertaken in early 1961.

Not surprisingly, the assessment commissioned by Gujarat showed that even larger storage could be provided at the Navagam site if the dam height was raised to what had been proposed earlier for the Harinphal site immediately upstream.

The command area and canal surveys also indicated the probability of a much larger irrigation potential with the higher dam and associated canal. More specifically, the studies conducted by the GOG found that a reservoir with FRL 400 feet or even greater would result in the optimal utilization of the untapped flow below the Punasa and increase irrigation benefits tremendously. Three sites around Navagam were explored for raising the height of the dam and high-level canal. Finally, in November 1963, the third site located upstream from Navagam recommended by the Geological Survey of India was selected and a dam height of FRL 425 feet was proposed.

But this project plan was not coordinated with the adjacent State of Madhya Pradesh, generating a set of inter-State conflicts that would bedevil development of the Narmada River Valley from that time on. In February 1961, the CWPC and the Government of Madhya Pradesh (GOMP) were also finalizing the Punasa Project Report, and it was decided that water from the Punasa Dam would irrigate a total of 3,200,000 acres. Subsequently, in 1963, GOMP revised the area it proposed would be irrigated from harnessing the Narmada River through the Puanasa Dam to 4,600,000 acres within its territory.[14] But the GOG's proposal for its revised Broach Project in 1963 projected use for approximately 4,000,000 acres of annual irrigation, up from the 90,000 acres approved by the Planning Commission in 1960.

An intense controversy arose over the share of waters to be utilized by these two riparian States, because the sum of the competing plans of Gujarat and Madhya Pradesh could not be sustained by the hydrology of the Narmada River. Federal-level authorities attempted but ultimately failed to moderate the emergent inter-State conflicts through a series of interventions over the next several years. This was partly due to the fact that that concern about water resources was a subject in India's federal system shared by both the Centre and the regional States. In addition, Nehru's death in 1963 resulted in a loss of legitimate power at the federal level, because no leader with his status was available to replace him as Prime Minister. Meanwhile, the States had become increasingly more aggressive in asserting their authority, particularly those States with Chief Ministers who had been strong figures in the Indian Independence Movement. This was particularly true of States such as Madhya Pradesh, Gujarat, and Maharashtra.[15]

The de facto decentralization of India's democratic political system thus limited the ability of federal intervention to mediate the inter-State controversies over the Narmada Projects. The Centre's Minister of Irrigation and Power, Dr. K. L. Rao, met with the Chief Ministers of the States of Madhya Pradesh and Gujarat at Bhopal, capital of Madhya Pradesh, in November 1963. Rao was a strong supporter of the idea that mega, multipurpose dam projects were in the national interest of India and had grand ideas for the Narmada River Valley. After overseeing the negotiations between the States, Rao announced a set of compromises from the meeting, called the Bhopal Agreement. The agreement stated that: (1) the Navagam Dam was to be constructed to a height of FRL 425 feet by the GOG and its entire benefits would be for that State; (2) the upstream Punasa Dam was to be built to a height of FRL 850 feet by the GOMP, the costs and power benefits of which were initially to be shared equally by all three States, but Madhya Pradesh was to repay Maharashtra's outlays within a twenty five year period; and (3) the Bargi Project was to be raised to a height of FRL 1390 feet in two stages by Madhya Pradesh with loan assistance from both Gujarat and Maharashtra.[16]

Despite the seemingly successful intervention of high-level federal authorities, the inter-State conflicts proved intractable. Satisfied with the outcome of the negotiations orchestrated by Minister Rao, the GOG Assembly speedily ratified the Bhopal Agreement and submitted a brief report to the CWPC outlining a Broach Project with a Navagam dam height of FRL 425 feet on February 14, 1964.[17] The GOMP, however, steadfastly refused to move ahead with the projects on the basis of the Bhopal Agreement, and plans to build big dams on the Narmada River were once again frustrated. In particular, the GOMP voiced strong objections to the conclusions reached at the meeting and argued that the Navagam Dam should not be raised higher than FRL 162 feet, which was the height of the river bed level at the Madhya Pradesh border. The GOMP Assembly refused to even consider ratifying the Bhopal Agreement.[18]

In another attempt to resolve the impasse, the federal Ministry of Irrigation and Power responded by appointing a high-level committee to formulate a development plan for the Narmada River Valley that would be acceptable to both these States in September of 1964. The Narmada Water Resources Development Committee (NWRDC) was headed by Dr. A. N. Khosla, the former chairman of the CWINC and later CWPC who had initiated the first investigations into developing the Narmada River Valley at the time of India's Independence, almost twenty years earlier. The committee's terms of reference included: (1) the formulation of a master plan for the integrated development of the valley with the assistance of the CWPC and the concerned States; (2) specific examination of the Navagam Dam Project and potential alternatives to it; (3) planning for a phased implementation period in order to produce maximum possible benefits; and (4) the development of proposals for the distribution of benefits and costs among the States.[19]

In other words, the mandate of the Khosla Committee was the same one that various governments and bureaucratic agencies had failed to fulfill since India had achieved its Independence. It was also a similar mandate that would continue to be given to various committees and groups with respect to the Narmada

Projects over the next thirty years. As the NWRDC proceeded with this work, it soon became apparent that the deliberations could not be completed in the four-month period proposed by the Ministry of Irrigation and Power, and its terms were extended a number of times. All the available project reports for the Narmada Valley were collected and the state governments provided additional documents, including draft master development plans of the basin. Numerous meetings were held with representatives of the State governments to discuss the various options available. The materials and responses received from the States showed that investigations for proposed projects had not even been undertaken in a majority of cases. Owing to the inadequacy of data, field studies and project reports, as well as continued inter-State wrangling, the NWRDC experienced considerable difficulty in coming to any firm conclusions.[20]

Nevertheless, the NWRDC ultimately formulated a set of recommendations based on a commitment to the intensive use of all the Narmada River Valley's water resources in the service of a top-down, technocratic vision of development as economic progress. No emphasis was placed on potential, long-term environmental impacts and very little on the social implications that would result from the priorities or plan formulated by the committee. Rather, the four guiding principles that were the basis for its proposals included: (1) National interests should have priority over State interests and the master plan for the Narmada Valley should thus generate the maximum amount of benefits possible to the country; (2) requirements of irrigation should have priority over those of power; (3) irrigation should be extended to the maximum area possible and, in particular, to the arid areas in Gujarat and Rajasthan; and (4) the quantity going to waste to the sea without providing irrigation or power should be kept to a minimum.

Irrigation was promoted over power generation because of continued problems with food shortages faced by India during this period and the heavy drain on foreign exchange involved in importing food grains to meet the shortfalls.[21] The move to irrigate arid areas along the India-Pakistan border was driven by national security considerations, as relations between the two countries had become even more tense.[22] As a result, a small portion of the Narmada waters was allocated to the neighboring State of Rajasthan for the first time, which added another dimension to subsequent controversies.

To "not let water go to waste in the sea" was extolled by the NWRDC. It was a slogan that would subsequently become a popular defense for proponents of big dam projects on the Narmada River during the 1980s and 1990s. Corresponding to these priorities of maximizing irrigation and power benefits for the country, the NWRDC submitted a master development plan to the Government of India on September 1, 1965, involving the construction of thirteen big dam projects in Madhya Pradesh and one in Gujarat. With respect to the Navagam Project in Gujarat, the committee advised that the dam should be built to FRL 500 feet, the height found to provide the maximum irrigation, power, and flood control and result in the minimum amount of water wasted to the sea. The water requirements of Madhya Pradesh, Gujarat, Maharashtra, and Rajasthan were assessed at 15.6 million acre-feet (MAF), 10.65 MAF, 0.10 MAF, and 0.25 MAF respectively. The amount of power to be generated was estimated at 2014 MW; it was to be shared by Madhya Pradesh, Gujarat, and Maharashtra in the ratio of 2.5:1:1. The NWRDC did not offer guidelines for sharing the costs of the proposed projects, although it initiated the first discussion of the need for an organized resettlement program for the people to be displaced from the development efforts.[23] But this issue was not made into a priority by the committee.

The inter-State conflicts were far from resolved by the NWRDC. While the deliberations of the committee were still in progress, the GOMP and the Government of Maharashtra (GOM) simultaneously entered into a separate set of negotiations contemplating the joint construction of a big dam at Jalsindhi solely for the purposes of power generation. The Jalsindhi Agreement, signed by the Chief Ministers on May 4, 1965, stated that the two States would "co-operate in the development of hydro-electric power at Jalsindhi on the Narmada River" at a location between the Harinphal and Navagam sites based on the master plan prepared by Madhya Pradesh. The GOM was to prepare investigations and estimates, and after being approved by the GOMP, would construct the project. The costs and benefits were to be shared by the States according to an agreed-upon, but arcane, formula. It was also stipulated that either State could propose modifications to the Jalsindhi Agreement at any time in the future.[24]

The nonbinding character of the Jalsindhi agreement clearly indicated that it was to be used as a tactic by the GOMP and the GOM in their bargaining with the federal government and the GOG. The NWRDC rejected the proposal for the Jalsindhi Project in favor of the more massive Navagam Dam/Broach Project and the set of other Narmada Projects it included in its master plan. The GOMP vehemently objected to the committee's response:

> In all fairness, if for some sound reason Navagam Dam must be built to such a height as would submerge one or more power houses proposed in Madhya Pradesh, the latter is entitled to receive from the power to be developed at Navagam . . . the full quantum of power that Madhya Pradesh would have generated in its own territory.[25]

Madhya Pradesh further claimed the right to use most of the water from the Narmada River originating in its borders, while offering to supply one-fifth of the net utilizable supply at Navagam to Gujarat by providing regulated releases from its upstream dams. Maharashtra advocated restricting the height of the Navagam Dam to FRL 210 feet. In effect, the views of the GOMP and the GOM supported the Jalsindhi Agreement they had reached earlier.

The federal Ministry of Irrigation and Power held discussions repeatedly over the next few years with the State governments to break the deadlock, ultimately to no avail. Meetings held in Delhi in the summer of 1966 resulted in agreements on a few important technical points such as the hydrology of the Narmada River, but wide differences over sharing of the waters, the areas to be irrigated in each State, and the specifications of the project at Navagam remained. Further meetings were held between the Chief Ministers of Gujarat and Madhya Pradesh in May, June, and December of the following year. These negotiations did not yield a mutually agreed upon solution to the disputes either; in fact, the differences seem to have widened.[26] By July of 1968, the GOG, left with no other option, petitioned for the appointment of a Tribunal to adjudicate the conflicts over the Narmada River, based on the 1956 Inter-State Water Disputes Act (see chapter 2). The federal Government, realizing that all other attempts at finding a negotiated compromise to the conflicts had failed, agreed to establish a tribunal.

THE LARGEST RIVER BASIN SCHEME PROPOSED IN INDIA

The Narmada Water Dispute Tribunal's proceedings and decisions of 1978/79 resulted in the formulation of arguably the largest river basin scheme ever proposed in India, including a set of big dam projects that would generate intensifying controversies and conflicts from the 1980s on. A range of actors central to these subsequent dynamics, such as the World Bank and other foreign donors, domestic and transnational nongovernmental organizations, as well as the people to be most negatively affected by the projects, was not involved in the Tribunal's deliberations. The subsequent struggles over the Narmada Projects from the 1980s on, as we shall see, often revolved around the lacunae in the NWDT award identified by these actors who were not historically included as participants in decision-making processes.

While the Tribunal's orders did determine many technical, financial, and institutional parameters of the Narmada Projects, central questions about benefits and costs, as well as resettlement and rehabilitation (R & R) of project affected peoples, were left unaddressed. In fact, the issue of R & R only emerged as a bargaining tool in the conflicts between the States—not as a central normative concern. Perhaps more importantly, no mention or debate about environmental impacts is discernible from any of the documents related to the

Tribunal's mandate, investigations, or final rulings. The Tribunal also did not consider alternatives to the construction of big dams for development in the Narmada River Valley. It was just assumed that these and associated projects were integral to the development, given the hegemonic vision and strategy in India at the time.[27]

The Narmada Water Disputes Tribunal (NWDT) took almost ten years before it issued its final award, primarily due to the continued wrangling among the Governments of the various States: Gujarat, Madhya Pradesh, Maharashtra, Rajasthan; and the federal Government of India, or Centre, which were parties to the matter. The federal Government constituted the Narmada Water Disputes Tribunal (NWDT) by Notification No. S.O. 4054 dated 6 October 1969, based on the Inter-State Waters Dispute Act passed by India's Parliament in 1956. The terms of reference of the NWDT included adjudication of matters relating to the apportionment of water from the Narmada between the riparian States, the height of Navagam or renamed Sardar Sarovar Dam, and submergence of lands.[28]

But the NWDT, far from moderating the tensions between the States, only provided a new arena in which the conflicts could be aired. The GOMP State filed a demurrer against the legality of the NWDT proceedings less than two months after the original orders to establish the Tribunal were given by the Government of India (GOI).[29] In the demurrer, the GOMP claimed that "the action of the Central Government in constituting a Tribunal is ultra vires, unconstitutional and void," because, "there was no water dispute within the meaning of Section 2(c) read with Section 3 of the 1956 Act and also that the Government of India had no material for forming the opinion that the water dispute could not be settled by negotiation within the meaning of Section 4 of the 1956 Act."[30] The GOMP further complained that the inclusion of the Government of Rajasthan's (GOR) claim for a share of the Narmada waters, which was based on that State's inclusion in the NWRDC's previously proposed master plan, would also be ultra-vires because it was not a riparian State. The GOM also questioned "whether the reference of the Central Government on behalf of Rajasthan was ultra-vires as no part of the territory of Rajasthan was located in the Narmada basin."[31]

The Sardar Sarovar Dam to be located in the State of Gujarat remained a focal point of the dispute. The demurrer suggested that the projects for the Narmada River proposed by GOMP and the GOM were only for purposes of power generation and therefore would not affect the interests of Gujarat State with respect to the flow and use of water into its territory. However, construction of the Sardar Sarovar Dam to the height of FRL 500+ feet, as was then being proposed by GOG, would submerge the three hydropower projects, particularly the Jalsindhi Dam, proposed by the other two states. In a separate petition, the GOMP pleaded further that "the Tribunal had no jurisdiction to give directions to Madhya Pradesh and Maharashtra to take steps by way of acquisition or otherwise for making submerged land available to Gujarat in order to execute the Navagam or Sardar Sarovar Project with FRL 530 or for rehabilitation of displaced persons."[32]

Like the NWRDC and other previous review committees, the NWDT, being a judicial body constituted by the GOI, attempted to impose a "National interest" above "State's interests" perspective on the proceedings. After hearing counsel from all the State Governments, as well as the Attorney General of the GOI, the Tribunal delivered its judgment over a year later on February 23, 1972. It agreed that the inclusion of Rajasthan in the dispute proceedings was ultra vires, because it was not a riparian State and therefore did not have the rights of a riparian State under the 1956 Inter-State Waters Dispute Act. It also held, however, that the constitution of the NWDT by the GOI was not ultra vires and that the proposed construction of the Navagam/Sardar Sarovar Dam involving submergence of lands in Maharashtra and Madhya Pradesh could form the subject matter of a "water dispute." More specifically, the Tribunal stated that "it had jurisdiction to give appropriate direction to Madhya Pradesh and Maharashtra to take steps by way of acquisition or otherwise for making submerged land available to Gujarat in order to enable it to execute the Sardar Sarovar Project and to give consequent directions to Gujarat and other party States regarding payment of compensation to Maharashtra and Madhya Pradesh, for giving them a share in the beneficial use of the Sardar Sarovar Dam, and for rehabilitation of displaced persons."[33]

In response to the Tribunal's rulings on preliminary issues, both the GOR and GOMP filed special leave petitions in the Supreme Court in May and June of 1972 and obtained a "stay to a limited extent" on the proceedings. The Supreme Court then ruled that the Tribunal proceedings should be suspended but that discovery, inspection, and other miscellaneous investigations of the Tribunal might go on. The Supreme Court also permitted the GOR to participate in these interlocutory proceedings.[34] At this time, five assessors specializing in hydrology, agriculture, civil engineering, and power engineering were appointed by the GOI to assist the Tribunal in its work. The absence of any social scientists or ecologists among this expert group offers evidence that social and, even more clearly, environmental issues—such as the impacts on tribal peoples or the potential loss of forest lands—were not prioritized.

The Tribunal's proceedings were stalled for all practical purposes, due to the writ petitions in the Supreme Court, until an agreement was reached among the Chief Ministers of the four States. In this agreement, signed on July 22, 1972,

Prime Minister Indira Gandhi was called on to not only allocate the shares of water from the Narmada River between the States but also to determine a suitable height for the Sardar Sarovar Dam.[35] A series of further meetings held over the next two years generated an agreement on July 12, 1974, brokered by Mrs. Gandhi. Although Mrs. Gandhi was by this time championing the cause of the environment, due to her experience at the 1972 United Nations Conference on the Environment, she did not take this opportunity to insert ecological issues into the development planning for the Narmada River Valley.[36]

But two crucial decisions were taken in this 1974 agreement brokered by the Prime Minister: (1) that the quantity of water in the Narmada River available be assessed at 28 million acre feet at a 75 percent reliability factor, and the Tribunal proceed on the basis of that assessment, and (2) that the requirements of Maharashtra and Rajasthan for use in their territories be 0.25 and 0.5 million acre feet respectively. The compromise that was reached by the Chief Ministers giving Rajasthan a portion of the Narmada River's waters effectively overturned the preliminary ruling of the Tribunal not to include that State in its deliberations. The agreement further directed the Tribunal (and not the Prime Minister, as asked for by the Chief Ministers in several of these areas), to determine the shares of Gujarat and Madhya Pradesh from the remaining net available quantity of 27.25 million acre feet of water, to determine the height of the Sardar Sarovar dam and canal level, and resulted in the withdrawal of the two special leave petitions filed earlier in the Supreme Court by the GOR and GOMP.[37]

Somewhat liberated by the 1974 agreement, the Tribunal conducted exhaustive investigations, traveled extensively through the river valley, evaluated other large water projects and water disputes' precedents from all over the world, and accumulated over 3,000 documents. Subsequently, the States and Centre were directed to examine in more depth the following areas: (1) hydrology of the river—even though the flow had already been set at 28 MAF by the 1974 agreement among the Chief Ministers, (2) areas to be submerged from the 30 big dam projects proposed, and (3) proposed rehabilitation measures—including cost estimates—for those to be displaced. As a result, over 81 studies were completed and filed.

Once again, not one of the studies directly or rigorously examined the environmental effects of the proposed projects for the Narmada basin. In addition, alternatives to big dams could not and were not investigated by the Tribunal. As Gujarat High Court Advocate Girish Patel later wrote, "The entire question of the proper use and development of the Narmada river along with the right to development of the people in the Narmada Valley, was in the political context of a federal structure, submerged into the bitter inter-State water disputes between Gujarat (with Rajasthan as its ally) and Madhya Pradesh and Maharashtra. . . . The very nature of the conflict between the States forced the Tribunal to choose or decide among the claims and counter-claims of the States, not on the basis of genuine development of the valley and real alternatives to major dams on the Narmada river."[38]

Resettlement of displaced people was not understood as a development goal and would not have even become an issue had it not been for its utility in the inter-State and federal-state wrangling over the shape of the master development plan for the entire Narmada River Valley. Displacement had become a more prominent issue, as Girish Patel later suggested, directly as a result of

> the rivalry among the contending States and expediency of politics at the time. Maharashtra and Madhya Pradesh were trying to put in as many obstacles in the way as possible. The rehabilitation issue was quite a potent tool in their hands. Gujarat State, equally determined and cynical if not more, wanted to remove these obstacles single-mindedly and at any cost. . . . It accepted total responsibility of settling all the oustee families of Maharashtra and M.P. in the command area of the SSP.[39]

When the offer was made, Gujarat officials had not even conducted any feasibility studies on this resettlement proposal and more than likely believed that they never would be held accountable to it. The GOMP and GOM never expected Gujarat to make such a move. But the GOG's proposal did set the stage for the Tribunal to issue a judgment that would be acceptable to all claimants.

The Tribunal's proposed scheme included the building of over 3,000 dams, including 165 big projects. It thus expanded tremendously on the ambitious proposals offered earlier by the various State governments and federal authorities. The two centerpiece projects of the plan were the FRL 455 foot high Sardar Sarovar Dam to be located near the border of Gujarat and the FRL 860 foot high Narmada Sagar Dam to be built upstream in Madhya Pradesh.

The Tribunal clearly attempted to find technical solutions, given the pressure to balance the competing political interests of the States. For example, FRL 455 feet was determined as the optimal dam height for the Sardar Sarovar Project, because it was ostensibly a technically sound option that would balance Gujarat and Rajasthan's interests in creating irrigation potential with Madhya Pradesh's and Maharashtra's interests in power generation. It also potentially reduced the amount of submergence—and therefore the corresponding extent of resettlement required—in Madhya Pradesh and Maharashtra from what would have occurred with the FRL 530 feet Sardar Sarovar dam proposed by Gujarat—without a great loss in total benefits.

Table 3.1 provides a snapshot of the relationship between the contentions of the GOG and GOMP with the corresponding decision of the Tribunal in a few

Table 3.1. Decisions of the Narmada Water Disputes Tribunal

	Madhya Pradesh	Gujarat	Tribunal
Area requiring water in Madhya Pradesh (100,000 acres)	70.70	30.00	68.05
Area requiring water in Gujarat (100,000 acres)	23.08	71.38	50.02
Water requirement of Madhya Pradesh (million acre-feet)	24.08	6.00	18.25
Water requirement of Gujarat (million acre-feet)	4.44	22.72	9.00
Total water requirement for both States (million acre-feet)	28.52	28.72	27.25
Full reservoir level Sardar Sarovar Dam (feet)	210	530	455
Full reservoir level Jalsindhi Dam (feet)	420	—	—

selected areas.[40] The massive discrepancy in the positions of the two States with regard to major parameters of the Narmada Projects is clearly shown. For example, the GOMP proposed a Sardar Sarovar Dam height of 210 feet and a Jalsindhi Dam height of 420 feet, while the GOG had argued for 530 feet for the former and not even building the latter. The GOG further proposed that Madhya Pradesh's water requirement was 6.00 MAF, while its own was 22.72 MAF; and the GOMP proposed that Gujarat's water need totaled 4.44 MAF, while its own was 24.08 MAF.

Critically, the one parameter which both Governments basically agreed upon was the total water requirement for both the States at a little over 28 MAF—a figure similar to the one agreed upon by the Chief Ministers in the 1974 compromise brokered by Prime Minister Indira Gandhi—a figure that would later become a central point of contention. Of the net 27.25 million acre feet of water ostensibly available to be distributed, the Tribunal allocated 18.25 million acre feet and 9 million acre feet respectively to Madhya Pradesh and Gujarat. The ratio of allocation among the four States were thus 73 for Madhya Pradesh to 36 for Gujarat to 2 for Rajasthan to 1 for Maharashtra, regardless of the quantum of water available in the Narmada River during any particular year. The Tribunal specifically mandated that the upstream Narmada Sagar Dam would be completed earlier to or simultaneously with the Sardar Sarovar Dam in order to provide regulated releases of water to the latter.[41] In terms of power generation, Madhya Pradesh received 57 percent, Maharashtra 27 percent, and Gujarat 16 percent. Various procedures for cost sharing with respect to the projects were stipulated.

The Tribunal also took strategic steps to bridge the competing claims of the States in the area of rehabilitation and resettlement. Yet, the fact that the Tribunal's orders only covered displaced peoples in Maharashtra and Madhya Pradesh and only from the Sardar Sarovar Project shows that the resettlement package was an incidental consequence of the broader disputes over the costs and benefits of alternative development plans proposed by the various governments for the Narmada River Valley. It was not a primary result of the concern for the human rights and welfare of the people to be negatively affected by the projects.

Thus, the GOMP and GOM were directed to acquire the lands to be submerged from the Sardar Sarovar Project for which Gujarat would compensate them monetarily. A process for notification and a timetable for resettlement were set up; entire villages with a wide range of civic amenities were ordered to be provided for the displaced; and once again a complicated scheme for cost-sharing was provided.[42] But no mention was made of the Gujarat oustees of the Sardar Sarovar Project nor the prospective negatively affected people from the dozens of other big dam projects that were included in the Tribunal's judgment.

The last series of Tribunal orders—directions setting up an elaborate machinery for the management of the proposed Narmada River Valley Projects—essentially reproduced the extant federal bureaucratic tensions that had created the stalemate in the first place. Each State was to independently execute the planning and implementation of the projects located in its territory, as well as to determine the pattern of use for the water allocated to it. So, for example, the Sardar Sarovar Construction Advisory Committee was entrusted by the GOG to supervise the Sardar Sarovar Project in Gujarat, and the Narmada Valley Development Agency was created by the GOMP to coordinate the Narmada Projects in Madhya Pradesh.

The ruling essentially created barriers to the basin-wide planning and the integrated river valley perspective that the Tribunal had itself utilized. It resulted in a bureaucratic framework that was highly competitive, with effective authority residing with the States. In an attempt to counteract this competitive institutional structure, an inter-State "Narmada Control Authority" was to be established consisting of seven engineers, one appointed by each of the respective States, and three members appointed by the federal Government to coordinate the separate State development activities in the Narmada Valley. The GOI was to appoint one of its three representatives as Chairman of the Authority with a tie-breaking vote at meetings when a decision affecting the interest of more than one State was at stake.

Centre-State and inter-State authority relations were thus internalized into a new set of bureaucratic and procedural structures that were to oversee the execution of the Narmada Projects. A Review Committee was mandated that could review the decisions of the Authority. But even its potential role was diluted. Decisions could only be taken at meetings in which the chairman and all the members were present. The Federal Minister of Irrigation was made chairman of the Review Committee and the Chief Ministers or their appointees of the respective States were to be permanent members. The Chairman was to act as the convener but had no voting rights. While the decisions of the Narmada Control Authority were to be binding on the respective States, the Review Committee was given the power to make decisions in the case that the Authority could not fulfill its duties or to override the Authority if necessary. But the means by which these rulings could be enforced was not stipulated.

The length of the investigations and proceedings, as well as the extensiveness of the Tribunal's award, would seem to have settled all matters. However, the States and the Centre sought further clarifications from the Tribunal, which delayed the finalization of the orders for another year. Specifically, the GOMP asked for the Sardar Sarovar Dam height to be reduced from 455 feet to 436 feet in order to reduce the numbers to be displaced within its territories; the GOG sought clarification regarding the time frame for completion of Narmada Sagar Dam and on the functioning of Review Committee; the GOM and the GOR asked for more specifics on the sharing of financial costs; and the GOI sought clarification on the effective functioning of the Narmada Control Authority.

A central issue raised, and one that would re-emerge again and again in the years to come, concerned the finality of the Tribunal's judgment. The Tribunal's responses to these challenges were included in the revised, final award and its publication in the Gazette of India on December 12, 1979. The NWDT judged that it could be asked for clarifications and explanations but not for re-adjudication of any issues, including the 455 feet height specification for the Sardar Sarovar Dam, and that the award was in fact final. Moreover, it was stipulated that a review of the award should be conducted in 2025, forty-five years after the award's publication, by an appropriately constituted judicial body. This NWDT final award set the stage for the increasingly dramatic struggles over the Narmada Valley Projects in the years to come.

INDEPENDENT DOMESTIC RESISTANCE PROVES INEFFECTIVE

Over the time that the Tribunal held its proceedings from 1969 to 1979, development dynamics had changed considerably. A novel range of domestic and transnational actors entered the fray, altering the trajectory of the Narmada Projects in profound ways.

The GOG, on the one hand, commenced its negotiations with foreign donors—primarily the World Bank, the United Nations, and Japan—for assistance with the Sardar Sarovar major dam component of the broader Narmada Projects in 1978, even before the Tribunal had given its final orders. An episode of domestic resistance against the Sardar Sarovar Project occurred soon thereafter in the State of Gujarat, but was ultimately unsuccessful because of the weakness of the tribal groups and local nongovernmental organizations involved.

On the other hand, the first phase of opposition by domestic groups in the State of Madhya Pradesh also was initiated in direct response to the Tribunal's orders, and the proposed Sardar Sarovar Project in particular—almost immediately after the rulings were made public. This initial grassroots mobilization in Madhya Pradesh also proved ineffective, since it was co-opted and later discarded by politicians caught up in the party/electoral exigencies of India's federal democracy.

The initial domestic mobilization in Madhya Pradesh against the Tribunal's ruling was sparked because the height of 455 feet for the Sardar Sarovar Dam specified by the Tribunal would cause submergence of prime agricultural lands in the Nimar district of the State. It quickly expanded and intensified because these lands were mostly owned by middle-class farmers with substantial material resources to deploy on social mobilization. In fact, a front-page story in the widely read *Times of India* newspaper was printed on Saturday, August 19, 1978—less than one week after the Tribunal's award was first handed down: "*Madhya Pradesh MLA on Fast: Award*—Cutting across party lines, reactions to the Narmada Tribunal award continue to be sharp. . . . A massive rally was organized by the 'Save Nimar' Committee yesterday in protest against the award."[43]

Then, on August 28, 1978, opposition party members created "virtual pandemonium in the vidhan sabha," or state legislative assembly, pressing the governing Janata Party to make a statement on the State of Madhya Pradesh's stand against the award. At the same time, "outside the assembly, hundreds of demonstrators from Nimar, the area which would be affected by the award, courted arrest under the all-party banner of Nimar bachao-sangarsh samiti," or "Save Nimar Action Committee."[44] At this point the Chief Minister of Madhya Pradesh agreed only that he and his cabinet would consider action against the Tribunal's decision on the height of the Sardar Sarovar Dam and assess the finality of the Tribunal's ruling.

The ruling Janata Party of Madhya Pradesh was compelled to petition the Tribunal to lower the height of the Sardar Sarovar Dam to 436 feet in order to minimize the areas to be submerged in Nimar, or risk its own survival in power.

Another, even larger, rally against the award occurred a little more than one week later on September 7th. During this protest a massive procession of the Save Nimar Action Committee marched through Bhopal, the capital of Madhya Pradesh. The procession assembled at the gates of Vidhan Sabha once again and, after forcing entry with the assistance of elephants and horses, was tear-gassed and lathi charged by the State police.[45] The police detained over 1,000 protesters; 365 were later jailed, and among them were Dr. Shankar Dayal Sharma (a future President of India), Mr. V. C. Shukla (a future Union Minister of Water Resources), and Mr. Arjun Singh (a future Chief Minister of Madhya Pradesh).[46] They were all members of the opposition Congress (I) Party in Madhya Pradesh at the time and utilized the backlash against the NWDT orders as a means to garner popular support for themselves and their political party.

But the degree and organization of resistance was not powerful enough to overcome the legal finality of the Tribunal's judgment and the political momentum it had engendered. As noted previously, the NWDT refused to reconsider this issue in its 1979 final award declaration, despite the GOMP's petition for redress. Rather, the NWDT stated that re-opening the hearings would have been against the procedures laid down by the Inter-State Water Disputes Act of 1956, which authorized only the clarification but not the re-adjudication of a Tribunal's decisions once promulgated. Partly due to the Janata's inability to secure a favorable ruling from the Tribunal, the Congress (I) won the 1980 election in Madhya Pradesh—under Arjun Singh, who, as a key opposition leader, had been extremely critical of the award in support of the people from Nimar.

The Save Nimar Action Committee withdrew its overt resistance believing that the Congress (I) Party leaders would continue their opposition to the Tribunal's specification of the Sardar Sarovar Project after being elected. They were disappointed, however, when Chief Minister Arjun Singh agreed to a political compromise and signed a tepid memorandum of understanding with the then Chief Minister of Gujarat Madhavsingh Solanki, also from the Congress(I), to "implement the decision of the Narmada Water Disputes Tribunal" as long as "both the States of Gujarat and Madhya Pradesh agree to explore the possibility of reducing the distress of the displaced persons as much as possible."[47] Arjun Singh and other Congress(I) leaders convinced the people from Nimar that they would be compensated for their lands, and the resistance ended.

As this brief episode of independent local resistance was occurring in Madhya Pradesh, the GOG moved quickly to initiate the Sardar Sarovar Project with the assistance of foreign donors. In November 1979, prior to the Tribunal's final award declaration, the GOG played host to the first reconnaissance mission of the World Bank, which had been watching the Tribunal's hearings with great interest. The description given in the report of mission at the time reiterated the historical context in which the projects had been planned and suggested that further, in-depth investigations had to be completed before the Sardar Sarovar Project could be supported by the Bank.

> When the NWDT (Narmada Water Disputes Tribunal) made its basic award in August 1978, the GOG (Government of Gujarat) had already prepared a fourteen-volume feasibility study for the SSP (Sardar Sarovar Project) and there was a strong desire to start building the project according to these plans. However, the existing plans had basically been prepared in the 1950's and 1960's . . . and were largely prepared against a background of riparian conflicts.[48]

The Bank's mission subsequently made three recommendations: (1) establishment of a high level Narmada Planning Group in Gujarat to coordinate the project; (2) the retention of domestic consultants to conduct various studies; and (3) the acquisition of foreign experts to investigate specific technical issues for which domestic consultants were not available. The World Bank secured $10 million from the United Nations Development Program to assist the GOG in implementing these recommendations.

Even though the Bank had made its skepticism of the quality and comprehensiveness of the existing studies and formulation of the projects clear, the GOG went ahead and submitted a project report without any further studies to the Central Water Commission (CWC) of the GOI for the techno-economic clearance that had been required of all big dam projects since the initial years of India's Independence. In fact, had the World Bank not been involved at this point, the version of the Sardar Sarovar Project that was submitted to the CWC by the GOG in 1980 would likely have been cleared relatively quickly.

The World Bank, under pressure from its critics, had already begun to initiate changes in its project clearance process by this time. The Bank's influence at this time was demonstrated by the fact that its recommendation to form a Narmada Planning Group was implemented prior to the final award of the NWDT, and thus well before the bureaucratic agencies prescribed by the Tribunal itself: the Narmada Control Authority and Narmada Review Committee were established approximately a year later on 10 October 1980. And, as former Secretary of Water Resources for the Government of India, Ramaswamy Iyer, pointed out, "World Bank involvement slows down the approval of projects like the Sardar Sarovar—because of the more rigorous appraisal and numerous conditions the Bank requires, these projects generally become gold-plated."[49]

Despite the fact that resettlement and rehabilitation (R & R)—particularly of indigenous or tribal peoples—and environmental concerns had begun to emerge as critical issues on the international and Indian development agendas during the 1970s, a preponderance of the initial studies after the Tribunal's ruling (sponsored by the Bank and the respective Indian authorities) focused primarily on technical and economic issues of the Sardar Sarovar Project. With the funds secured by the Bank from the UNDP, the Narmada Planning Group hired consultants to conduct studies required for domestic clearance of the project and also for additional international financing. This included the Tata Economic Consultancy Services of Bombay, Maharashtra, which calculated the benefit/cost ratio for the Sardar Sarovar Project to be 1.84 at economic prices and 1.39 at market prices, well above the level needed to make it viable.[50] A few studies were conducted on R & R by the Center for Social Studies in Gujarat; however, not a single one was commissioned to investigate environmental impacts.

The Tribunal's ruling on R & R had mandated land and not just cash compensation for each family, continued support through grants and financial assistance, and relocation of communities as units to new villages. Yet, three problems quickly emerged. First, the Tribunal had no oversight or authority over the implementation of the award, and the bureaucratic machinery that was stipulated to execute the projects depended ultimately on the cooperation among the State and federal Governments—in particular the various agencies in charge of water and power resources. Second, the question of what to do with encroachers on State-owned forest and wastelands (of whom many were tribal peoples who did not have land titles) was not explicitly addressed. Finally, the award specified the R & R package for the displaced of Maharashtra and Madhya Pradesh, and for the GOG's assistance to those States, but made no mention of the package to be offered to those to be displaced from the Sardar Sarovar Project located in Gujarat itself.

In fact, further demonstrating the relative ineffectiveness of indigenous peoples and human rights norms in India at the time, the GOG issued a government resolution on June 11, 1979, which specified that only landholding oustees—that is, those with legal land titles—in that State would receive land as compensation.[51] The Government Resolution did not even mention the Tribunal award's more expansive policy for Madhya Pradesh and Maharashtra. This meant ruin for most of the tribal families to be displaced by the Sardar Sarovar Project in Gujarat, because they tended to have joint landholdings, rarely had land titles as a result of their past marginalization, or were simply encroachers.

This 1979 Gujarat Government Resolution was not reformed until sustained resistance by the Gujarat oustees and transnationally allied pressure from nongovernmental organizations inside and outside India compelled such change several years later. But prior to that, the second episode of independent domestic resistance to the Sardar Sarovar Project occurred in the early 1980's in opposition to the 1979 GOG resolution on resettlement and rehabilitation. As a report of the events at the time recounts, "oustees from the first five villages to be displaced after the award from the construction of rock fill dykes were offered the minimum of 5 acres of worthless government wastelands more than 100 kilometers away—away from their society and culture."[52] In response, the first grassroots mobilization in Gujarat, after the protests in Madhya Pradesh against the award had already died down, was launched in March of 1983 by the oustees of these five villages.

These displaced peoples were assisted by nongovernmental organizations that had been working to empower tribal peoples in the area, such as the Rajpipla Social Services Society and Chatna Yuva Sangharsh Vahini, since the mid to late 1970s.[53] These domestic nongovernmental organizations had taken on the task of organizing the project-affected peoples to defend their rights in the hopes of improving the compensation packages that would be given by the GOG. They also conducted research on the inequitable development and land distribution patterns in the villages, publicized these issues through the press, and lobbied authorities and sympathetic political elites through correspondence domestically within India.[54]

Indeed, R & R was increasingly becoming a priority issue in Gujarat, and in India more broadly, due to the criticism by nongovernmental organizations and mobilization by various groups around the country on this issue. At the same time, the subject was increasingly being framed in indigenous peoples' and human rights norms outside of India's borders with respect to big dam projects specifically, as well as development more generally (see chapter 6).[55] The World Bank, for example, had issued a statement on involuntary resettlement by February of 1980 stating that, "the Bank's general policy is to help the borrower to ensure that after a reasonable transition period, the displaced people regain at least their previous standard of living and that so far as possible, they be economically and socially integrated into the host communities."[56] This operational statement was updated two years later with a specific focus on tribal peoples: "As a general policy, the Bank will not assist development projects that knowingly involve encroachment on traditional territories being used or occupied by tribal people, unless adequate safeguards are provided."[57]

Despite the increasing institutionalization of norms on resettlement and indigenous/tribal peoples at the Bank, practice did not always follow policy reforms. As Michael Cernea wrote on the World Bank experience with the Narmada Projects: "During 1982–83, four Bank missions, two each for pre-appraisal and appraisal were mounted, but none of them appraised the resettlement component (of the Sardar Sarovar Project)."[58] Subsequently, during the month of July 1983, the Bank sent three letters to the federal and State govern-

ments in India focusing on the need for comprehensive resettlement and rehabilitation plans to be formulated, but never followed up on their requests.

The independent monitoring efforts of domestic nongovernmental organizations, moreover, had marginal impact on domestic authorities at the time.
The Bank received a letter from Dr. Anil Patel of Arch Vahini—one of the Gujarat nongovernmental organizations—in August 1983 detailing the problems
they had discovered on the ground with respect to resettlement.[59] Due to the
continuing unresponsiveness of the various Indian Governments and growing
apprehensions of some Bank staff as a result of Patel's letter, the World Bank decided to send an independent consultant to investigate the social impacts of the
Sardar Sarovar Project in particular and the capacity of domestic officials and
project authorities to mitigate them.[60]

As a result of continued domestic monitoring over the next six months, the
World Bank more vigorously urged the various Indian Governments to improve
their policies on resettlement. However, as Bradford Morse and Thomas Berger,
heads of the World Bank's 1992 Independent Review Team for the projects, later
noted, "Indian reaction . . . was defensive. Letters were sent to the Bank from the
Government of India stating that 'necessary steps are being taken to formulate
a rehabilitation plan,' and that no Bank mission 'should be mounted specifically
for this purpose.'"[61] Nevertheless Thayer Scudder, a renowned expert on issues
of resettlement from the California Institute of Technology, was contracted to
lead a resettlement appraisal mission to India in September 1983. Scudder returned with a highly critical report that echoed much of what Anil Patel of Arch
Vahini had documented in his letter to the Bank. He concluded that resettlement was likely to take place under highly unfavorable circumstances given the
paucity of information available, the inadequate institutional frameworks in the
three States, the noninclusion of encroachers in the Tribunal's award, and the lack
of a comprehensive rehabilitation policy at the all-India level.[62]

Responding to the prodding from the World Bank in April 1984, the Narmada
Control Authority produced a report in which an outline resettlement plan was
formulated.[63] The GOG also passed a new Government Resolution establishing
a committee to look into the purchasing of private lands for resettlement purposes, and separately contracted with the Centre for Social Studies in Surat, Gujarat, to prepare a detailed socioeconomic survey of the Gujarat oustees.

At the same time, the process of organized mobilization at the grassroots
gathered greater momentum. A massive march to Kevadia Colony—the construction headquarters near the Sardar Sarovar Dam site—involving oustees
from all the villages to be submerged in Gujarat and joined by potentially affected villagers from nine villages of Maharashtra, occurred on March 8, 1984.
The villagers demanded five acres of compensation for each family to be dis-

placed, including major sons and encroachers, and full implementation of the NWDT's package for Gujarat oustees. These demands were responded to by Amarsinh Chaudary, the Gujarat State Minister of Irrigation and Tribal Affairs at the time and later Chief Minister, with verbal assurances that the rehabilitation policy would be improved.

Yet the promises given to the people and domestic nongovernmental organizations that their demands would be met by the GOG generally remained unfulfilled. The struggle took on various forms including a Rasta Roko to stop work on construction of the rock fill dykes, as well as the first writ petitions submitted by domestic nongovernmental actors to the Gujarat High Court and to the Supreme Court of India against the Sardar Sarovar Project. Anil Patel's comments on the initial effects of the domestic-level social mobilization in Gujarat are revealing: "The Gujarat Government was rattled because we were there when construction just began, and they thought they were finally clear after the Tribunal award. But our experience early on was not very encouraging; nobody ever tried to repress us, nothing like that, but government officials thought we could be bought off."[64] In fact, the various tactics employed by the domestic opposition did not have much of an impact until a transnationally coordinated lobbying campaign targeted at the World Bank's involvement with the SSP supported by foreign nongovernmental organizations was organized.

EMERGENCE OF THE TRANSNATIONAL RESETTLEMENT REFORM CAMPAIGN

Within a short period of time, an initial transnational coalition attempting to reform the R & R—but not environmental—aspects of the Sardar Sarovar major dam component of the Narmada Projects emerged and was somewhat more effective. This transnational coalition consisted primarily of domestic nongovernmental organizations working with tribal groups from Gujarat State that constituted the second grassroots opposition discussed previously, linked with transnational nongovernmental organizations working primarily in the areas of indigenous peoples and human rights. India's democratic political system, as will be evident, continued to shape the trajectory of the Narmada River Valley Projects by conditioning the activities of these actors and the success they achieved in reforming the Sardar Sarovar Project.

From 1984 on, the federal and State governments in India as well as the World Bank began to face increased pressure from above by foreign and international nongovernmental organizations, as well as from below by grassroots groups and domestic nongovernmental organizations, and the increasing transnational links between the two. Arch-Vahini was being funded to conduct health-related work in the tribal areas of Gujarat by Oxfam, the transnational nongovernmental organization headquartered in the United Kingdom that had historically been involved in funding local, small-scale development projects in the third world. By the early 1980s, however, Oxfam had established a new policy unit within its organization to conduct activist work and was gradually initiating transnational lobbying efforts on development issues. John Clark, head of the new policy unit, believed that a campaign to reform the Sardar Sarovar Project would help Oxfam make a broader impact on development practices around the world and, in particular, on the policies of the World Bank.

Links with nongovernmental organizations working at the international level empowered the Gujarat-based domestic opposition in a number of ways. These transnationally allied groups provided: (1) greater access to the World Bank, (2) increased pressure on Indian authorities "from above" via their lobbying of the World Bank and their own governments; and (3) critical information that could be used to pressure Indian authorities "from below." Anil Patel later remarked that although initially Arch Vahini was hesitant to get involved in a transnational campaign on the resettlement issue, the organization quickly realized that the GOG in particular would be more likely to reform and implement better policies for resettlement and rehabilitation if the Bank was pushed to act.

> John Clark of Oxfam's policy unit came to know about the Sardar Sarovar Project and he said to me, "if you agree, I'll start raising this issue in the U.K. and at the World Bank." At first I said, "No, no, that will just create problems for us with the Government." But by 1984 we knew that the Gujarat Government was playing with us. This was a crucial time when the R&R policy was being shaped and we enlisted Oxfam's help because at that time the World Bank would not take any documents from us—we were just a small, local nongovernmental organization but when John Clark of Oxfam wrote to the World Bank, they had to pay attention to him.[65]

Domestic nongovernmental organizations, however, were also a critical source of information for their foreign counterparts and legitimized the activities of the latter as knowledgeable advocates of the people to be affected. After the Gujarat nongovernmental organizations agreed to participate in the campaign, John Clark and Oxfam moved quickly. He sent correspondence to the World Bank stating Oxfam's apprehensions about the resettlement component of the Sardar Sarovar Project, and enlisted the support of individuals in key positions in a variety of different organizations. Soon a transnational coalition was set up among Oxfam, Survival International, The Ecologist, Arch Vahini, the civil rights nongovernmental organization Lok Adikar Sangh in Gujarat, the domestic environmental nongovernmental organization Kalpavriksh in New Delhi, and others.

The initial lobbying and advocacy efforts by this transnational coalition of actors motivated the World Bank to send a post-appraisal mission to India on the resettlement component of the Sardar Sarovar Project in August 1984, led again by Thayer Scudder. Scudder met Anil Patel of Arch-Vahini in person for the first time while he was in Gujarat and told him, "Tell me whatever you want, give me in writing whatever you need because there is going to be a loan agreement with the World Bank and all the Governments in India, and I am going to write that part of the agreement." He also met and exchanged information with Medha Patkar, the later leader of the "Narmada Bachao Andolan" or "Save the Narmada Movement," who had just begun her work with Maharashtra tribals in 1984, during the same trip.[66] Scudder prepared an Aide Memoire to the Bank shortly after leading this post-appraisal mission, reiterating that, "In order to comply with Bank/IDA policy, the Government of India, the Narmada Control Authority and the States concerned would be required to provide at negotiation an overall detailed plan for the resettlement and rehabilitation of the oustees."[67]

A set of dynamics had now been established by which domestic and external nongovernmental organizations monitored the actions of the World Bank and Indian governments, holding them accountable to globalizing norms, in this case, on the human rights of tribal or indigenous peoples as well as resettlement. Soon thereafter, negotiations for a credit and loan agreement were held between the Bank staff and an Indian delegation led by Mr. A. Thappan of the federal Government of India's Ministry of Finance in November 1984 and January 1985 and completed on 1 February 1985. At the time of the negotiations, Survival International, a human rights nongovernmental organization established to defend the rights of indigenous peoples around the world based in London, took an especially strong position because of the thousands of tribals, or adivasis as they are called in India, to be displaced by the Sardar Sarovar Project. In a letter sent to the Bank and to the International Labor Organization (ILO) on January 28, 1985, Survival International stated its concerns and the consequences of the Bank's failure to ensure proper resettlement:

We should like to remind you that this resettlement programme, besides being inhumane and liable to result in serious physical and social problems for the tribals involved, is also illegal. India is signatory of ILO Convention 107 which not only acknowledges the right of tribal populations to the lands that they traditionally occupy (Article 11), but also, in the exceptional instance that they be removed from their lands, their right to the provision of lands at least equal to that of the lands previously occupied by them, suitable for their present needs and future development (Article 12) . . . We look forward to hearing how the World Bank intends to deal with this issue and to receiving details of the resettlement programme once it has been agreed upon.[68]

This was the first time that the Government of India's potential violation of ILO Convention 107 with respect to the Sardar Sarovar-Narmada River Valley Projects was openly referred to, although it would not be the last. Four days earlier, Thayer Scudder had also sent a letter to Ronald P. Brigish of the World Bank to pass on the worries that Anil Patel of Arch Vahini had conveyed through his correspondence to him about the GOG's duplicity with respect to the issue of R & R.[69]

These transnational lobbying and advocacy efforts gradually began to have an effect on the World Bank. A President's Report and Recommendation of the World Bank entitled "Narmada River Development (Gujarat) Water Delivery and Drainage Project," based on a Staff Appraisal Report of the Sardar Sarovar Project, was submitted to the Bank's Executive Directors on February 13, 1985. The report surprisingly stated that the Bank's involvement thus far had been instrumental in "the formulation of a comprehensive and equitable resettlement and rehabilitation program for oustees."[70] Tim Lancaster, the United Kingdom's Executive Director, armed with the information he had received through the transnational coalition from Arch Vahini via Oxfam and Survival International, strongly criticized this view about the resettlement component of the agreement at the Board's meetings held between March 5th and 7th in Washington, D.C. He forcefully argued that the Bank and Indian authorities had much to do in order to fulfill their obligations based on emergent global norms and procedures, including ILO Covenant 107, and the Bank's own policy platform.

Consequently, the Board of Directors approved the loan and credit agreements of $450 million on the condition that the respective Indian State Governments ensured that they would execute the program that had been agreed upon in the negotiations, as well as adopted and implemented resettlement policies in the future that would satisfy the Bank's principles and guidelines. A response from the World Bank to Survival International, dated April 19, 1985, provided an outline of the resettlement program that had been agreed upon:

> . . . a resettlement and rehabilitation (R&R) program has been negotiated between the Bank and Governments of India, Gujarat, Madhya Pradesh, and Maharashtra. The program forms an integral part of the project and has been incorporated in the legal agreements. . . . We believe that the agreed upon R&R program is broad-ranging and comprehensive. . . . The R&R plan includes provision to ensure the participation of all oustees. In order to ensure that the program is implemented in an equitable and timely manner, it will be carefully monitored and evaluated by independent agencies. . . . The Bank will finance monitoring and evaluation activities under the project. The GOI will be responsible for the overall monitoring and evaluation of the entire R&R effort and will report to the Bank semi-annually and annually on the progress of the program. . . .[71]

The R&R plan proposed by the Bank was primarily based on the 1979 NWDT award but added three further elements reflective of globalizing human rights norms on resettlement and indigenous/tribal peoples. First, it proposed that the GOI would release forest lands to facilitate land-based resettlement and second, if this was not possible, alternative means of livelihood would be offered to ensure that all oustees would "improve or at least regain the standard of living they were enjoying prior to their displacement."[72] Finally, the agreements supposedly also provided that encroachers would be given land for the lands they had been cultivating, even if they did not have land titles. All three of these would soon prove problematic and highly controversial.

During this time, all actions taken by the Bank and Indian authorities were being vigorously monitored by transnationally allied nongovernmental organizations. The loan and credit agreements were signed with the federal government of India and project agreements with the State governments on May 10, 1985. Survival International immediately sent a critical response to the World Bank based on the information it was receiving from the Gujarat nongovernmental organizations, Oxfam and other members of the transnational coalition. On May 23, 1985, President Robin Hanbury Tenison of Survival conveyed the following message to C. L. Robless, Chief of the India Division in the South Asia Programs Department of the World Bank, sending copies of the letter to the British, Japanese, West German, French, United States, and Indian Executive Directors of the World Bank:

> Thank you for your letter of April responding to the concerns we expressed regarding the tribal peoples to be affected by the Narmada Dam Project. . . . We have been very pleased to learn that an agreement has been negotiated between the Bank and the Governments. . . . Your letter does not however address a number of key points of my letter of 28 January. In particular I would like to draw your attention once again to the great importance of ensuring that the tribals who are to be relocated to make way for the dam project are compensated for the loss of their traditional lands (not just those that happen to be under cultivation or to which the tribals have managed to gain title). . . .

The letter further stated that based on the information that Survival International received from nongovernmental organizations in India, the GOG was ignoring the legal agreements signed with the Bank and was unlikely to compensate oustees that were encroachers. The World Bank responded by reminding the GOG of its legal and ethical obligations and, as a result, the Gujarat legislative assembly passed a resolution entitling encroachers to a minimum of 3 acres and maximum of 5 acres of land.[73]

Domestic lobbying and advocacy efforts had also intensified during the same period of time. The Gujarat nongovernmental organization's public interest litigation on behalf of the tribal oustees of Gujarat had reached the Supreme Court of India. The Indian courts had begun to play a critical role in the ability of the domestic and international opponents to monitor and pressure Indian authorities. They also provided means by which various civil and human rights violations perpetrated by proponents of the projects were reviewed and punished.

In fact, without the relatively autonomous functioning of the courts at both the federal and State levels of India's democratic system, it is highly unlikely that the transnational allied advocacy with respect to the Sardar Sarovar Project would have ultimately been as effective. In the proceedings of the first Supreme Court case in 1985, as had been related to the World Bank by Survival International, the GOG had clearly stated that encroachers were on the land illegally and therefore had no rights to it. While trying to get a copy of the World Bank loan and credit agreements to verify GOG's actual acceptance of the conditions contained in them, Anil Patel and Arch-Vahini had also sent their Supreme Court affidavits to Oxfam who forwarded them to the United Kingdom's Executive Director to the World Bank as well as to Survival International.

> In May 1985, the loan agreement was signed and in it there was an operative clause that oustees will regain the previous standard of living from which it was understood that encroachers would be given land. By then, we knew also that the GOG was up to mischief and they were just going to throw these people out. We were desperate to get a copy of the loan agreement but the Gujarat Government would not give it to us. So we went to the Supreme Court in 1985 where the GOG stated that encroachers were on the land illegally and therefore had no rights. Scudder had told us that they were to be included but the rest of the World Bank would take no cognizance of us because we were not an elected authority. So we sent our affidavits to John Clark of Oxfam, U.K., and he would write letters to U.K. Executive Director, Tim Lancaster, and he raised a stink at the Board of Executive Directors. Lancaster wanted to know what was in the World Bank loan agreement and he wrote letters giving information to Oxfam and Survival International that was sent to us sometime in 1986 and I was shocked. In those days, the Gujarat Government was just lying to the Supreme Court. Even John Clark couldn't understand the double game Gujarat was playing—agreeing to give encroachers land to World Bank while denying their rights in the Supreme Court.[74]

Based on the arguments presented at the hearing, India's Supreme Court issued an interim stay order on construction of the rock filled dykes and appointed an inquiry commission to investigate the matter further. The commission filed its report on July 11, 1985, criticizing the resettlement performance to date and indicting the GOG for illegal occupations of private lands as well as willful violation of the Supreme Court's stay order. However, while the GOG was denying the rights of encroachers in the Supreme Court of India, not implementing its own resolutions on rehabilitation among those displaced from the rock-filled dykes and moving ahead with construction on the project, it continued to agree to the World Bank's condition to give land to encroachers and implement a more comprehensive resettlement policy.

Under the mounting pressure from the transnationally allied advocacy coalition, the World Bank again sent a letter reminding the GOG that fulfillment of the resettlement program was a condition of the loan and credit agreements it had signed with the Bank. Once again, the GOG modified slightly its resettlement policy by another Government Resolution on November 1, 1985, in which it conceded to apply the NWDT principles to the oustees who had land titles in Gujarat and offer a choice of lands as per the Bank's recommendation.[75]

But the policy still did not meet emergent global norms on human rights and tribal peoples because it excluded major sons and encroachers and stated that oustees would forfeit their claim to land if they did not accept one of the three options given to them. Since many of the lands offered were waste or un-irrigated lands, the policy revision amounted to very little. In December 1985, the first monitoring and evaluation report of the Centre for Social Studies, Gujarat—that had been mandated by the World Bank as part of its loan package— was completed. The report once again highlighted the particular problems of tribal encroachers and the lack of information available on the negative social effects of the Sardar Sarovar Project. The Supreme Court in India passed an order shortly thereafter requiring that a displaced person, "shall be provided either alternative land of equal quality but not exceeding three acres in area and if that is not possible, then alternative employment where he would be assured a minimum wage."[76] As a result, the GOG passed a Government Resolution increasing the loan subsidy levels for oustees.[77]

A further tactic of transnationally allied advocacy efforts to pressure Indian authorities to comply with globalizing norms and principles was also being applied during this time. On October 23, 1985, the International Federation of Plantation, Agricultural, and Allied Workers sent a letter communicating information and criticisms based on ILO Convention 107 it had received from Survival International with respect to the Sardar Sarovar Project to the ILO's Committee of Experts. These comments were forwarded to both the Government of India and the World Bank in a clear attempt to hold these actors accountable to the convention. The Bank, in its response dated December 23, 1985, stated that "the Bank's policy and arrangements on this project are in no way inconsistent with, nor do they fall short of, the provisions of" the convention. The Government of India, in its reply of March 10, 1986, stated that the concerned governments were "fully aware of and committed to the present Convention, and that while implementing the resettlement and rehabilitation programme, the Government and Project Authorities will ensure that the Convention is not violated," and that the "State governments are providing rehabilitation for the landless by the provision of other means of assured livelihood, though not necessarily by ownership of land in the new plots."[78]

But the ongoing inquiries of the ILO Committee of Experts added one more barrier to the World Bank and Indian authorities in implementing the Sardar Sarovar Project. In April 1986, at its annual meeting, the Committee of Experts reviewed the case and "raised a number of questions related particularly to whether different categories of displaced tribal persons were being compensated for the loss of their lands in a manner consistent with the Convention's requirements," and requested the Government of India to provide detailed information on the manner in which the conditions stipulated in the World Bank agreements were being implemented.[79] The Government of India did not send a response assuring that the conditions were being met until late in 1987 but this correspondence involving Survival International, organizations like the International Federation of Plantation, Agricultural and Allied Workers, the ILO's Committee of Experts, and the Government of India on compliance with Convention 107 would continue for a number of years to come.

The changing development dynamics around big dams shaped by the domestic institutionalization in India of globalizing norms on the environment further altered the context in which the Narmada Projects had to be built. For example, in April of 1986, the federal government's Ministry of Environment and Forests (MOEF), after reviewing the environmental impact assessments of the GOG and GOMP, issued a confidential note to Prime Minister Rajiv Gandhi. The MOEF voiced its disapproval of the Sardar Sarovar and Narmada Sagar components of the Narmada Projects, partially because "resettlement capability surveys of land identified had not been done," and the likelihood of adequate resettlement seemed low.[80] This criticism with respect to resettlement was connected to the increasing divergences between the MOEF on the one hand, and the federal Government of India's Ministry of Water Resources and the State governments on the other, over the environmental impacts of the projects.

As the next chapter will more clearly demonstrate, there was a progressive integration of human rights, indigenous peoples, and environmental concerns that strengthened the second transnational coalition that was coalescing to not reform but halt the implementation of the Sardar Sarovar Project altogether.

Indeed, during this period of time, Medha Patkar and a few other activists established the Narmada Dharangrasta Samiti, or Narmada Action Committee, a nongovernmental organization dedicated to mobilizing the tribals of Maharashtra who were also to be displaced by the project. Patkar had been working in the Narmada River Valley investigating the issues for more than one year when she met William Partridge and Abdul Salam of the World Bank who were there on the first bi-annual Bank supervision mission of the Sardar Sarovar Project in 1986. In her words, "Partridge and Salam were very sympathetic to the plight of the tribals in the Valley."[81] In fact, the report they submitted to the World Bank on July 12, 1986 was extremely critical of the GOG's implementation of resettlement policies, observing that if changes were not soon made, the principles of the Bank's own policy of 1980 would be violated.[82] Furthermore, beginning in late 1986 and 1987, the Multiple Action Research Group—a nongovernmental organization based in Delhi—published the first of a series of reports entitled "Sardar Sarovar Oustees: What do they know?" focusing on the plight of the people to be displaced in Madhya Pradesh.

As a result of this broadening of domestic opposition and the continuing transnationally allied criticism at the international level, there were now increasing concerns from Bank staff not only about the GOG's resettlement policy, but also about the GOM's and the GOMP's resettlement policies, because all of the three state governments had signed the 1986 loan and credit agreements with the Bank. As Bradford Morse and Thomas Berger later recounted, "In April 1987, in response to mounting pressure to examine what was happening to implementation of resettlement policies for these Projects, the Bank sent its largest ever mission to the region."[83] Its central task was to focus especially on the status of the tribal oustees. The Bank discovered that the negotiated condition to provide each oustee with at least five acres of land was not being implemented by Indian authorities.

The mounting pressure being applied on Indian authorities by transnationally allied nongovernmental organizations via the World Bank is vividly demonstrated by the following incident recounted by Anil Patel of Arch Vahini:

The World Bank lawyer, Escadero, who knew what was in the agreement because he had written it, and he was very sharp—like an Indian he was—he came in April, 1987 to check what was going on. He got a hold of our affidavits to the Supreme Court from our allies and read them and he confirmed that Gujarat was lying and then he called us to meet him at Vadgam village. He said to me, "Anil, I want to talk to these people. I want to give a speech and I want you to translate it because I don't trust that they will give a correct translation," and he was pointing to the GOG representatives, "to tell them what the World Bank has in mind." He then gave a long speech, which I translated, and said to the tribals "whatever the GOG says, each of you is entitled to five acres of land" and the GOG representatives were so embarrassed, but the tribals were very pleased.[84]

Two post-mission letters to the GOG set a ten-month deadline to remedy this failure of compliance with resettlement conditionalities and implicitly warned that the loan and credit funds might otherwise be suspended.[85]

But tensions in the Narmada River Valley were rising as there were now stirrings of mobilization in Madhya Pradesh—initially again in Nimar, the sight of the first grassroots resistance against the Tribunal award in 1978 discussed early in this chapter—in addition to the progressively more organized opposition in Gujarat and Maharashtra. In November 1987, the newly appointed President of the World Bank, Barbar Conable, visited New Delhi and met with representatives of domestic nongovernmental organizations working with the oustees. While the meeting seemed mostly to be a public relations exercise, the mission that accompanied Conable reiterated the problem of noncompliance with respect to the two hectares of land for oustees.

Caught between the determined and sustained resistance of the domestic groups at the grassroots and the Supreme Court case at the federal level, on the one hand, and the massive transnational campaign of lobbying with the World Bank, on the other, the GOG began to grudgingly make one after another concession on policy. On the last day of the Bank's Winter 1987 mission, the GOG initially issued a Government Resolution affirming the two hectare minimum for landed oustees. Two weeks later, on December 14th and 17th, three more resolutions were passed.[86] The first resolution granted that landed oustees, who were eligible for a minimum of two hectares of land, would be given an ex gratia payment equal to the difference between the compensation received for a submerged land-holding and the market price of buying five acres of land.[87] The second and third resolutions increased the subsistence allowances and grants to be given to the oustees during the process of relocation.[88] The final resolution—and by far the most unexpected—completely reformed the resettlement policy by adopting the same package for all landless oustees of Gujarat, not just those cultivating unauthorized lands, that was to be given to landholding oustees.[89]

SUPPORTING IMPLEMENTATION OF REFORMS VS. OPPOSING THE PROJECTS

The initial transnational coalition to reform the resettlement aspects of the Sardar Sarovar Project had thus won a clear and major victory after almost five years of mobilization and lobbying at various levels of governance from the local to the international levels. While two earlier independent attempts at domestic resistance failed to make a sustained impact on Indian authorities, with the forging of transnational links to international human rights and indigenous people's organizations, the balance of power between the proponents and opponents of the projects was substantially altered. The global spread of norms and procedures on resettlement and rehabilitation of displaced peoples based on human and indigenous peoples' rights such as those established at the World Bank were critical to the empowerment of this transnational reform coalition. Finally, India's democratic opportunity structure—in particular the ability of citizens to organize themselves, to form links to external actors, to have access to information to pressure and hold authorities accountable through various institutional mechanisms, and especially the courts—facilitated the effectiveness of the resettlement campaign.

But a rift soon emerged between those who took a "no implementation of resettlement, no dam stand" and those who progressively took a more strongly anti-dam position regardless of the resettlement reforms. While the former groups, such as Arch-Vahini and Anand Niketan Ashram began to cooperate with project authorities on implementation of the policy reforms, other groups such as the Narmada Dharangrast Samiti and Kalpavriksh supported by various foreign and international human rights and particularly environmental nongovernmental organizations took a stand that a complete and comprehensive re-evaluation of the Narmada Projects was still required before any further construction was acceptable. The latter groups combined a number of critiques: that the resettlement reforms were not likely to be implemented, that the projects would cause irreparable environmental damage, and that the projects' economic costs far outweighed their economic benefits. They thus launched an even more massive multilevel, transnational campaign against the Narmada Projects as both a concrete and symbolic example of a destructive vision of development that had to be stopped, no matter what the cost.[90]

❖

CHAPTER FOUR

The Transnational Campaign to Save India's Narmada River

The rapidly changing context for big dam building and development since the 1970s—especially the global rise, spread, and deepening of norms pertaining to environmental protection, human rights, and indigenous peoples—was critical in stalling the construction of the Narmada Projects and the Sardar Sarovar major dam in particular during the 1980s and 1990s. Yet, while the global institutionalization of these norms was clearly a facilitating factor, it could not have produced this outcome independently of the formation of a second transnational coalition, consisting of nongovernmental organizations from around the world linked to a well-organized domestic social movement, that was capable of using these norms in a massive, multilevel campaign against the projects. The sustained grassroots mobilization by this domestic social movement and the existence of democratic institutions within India itself greatly contributed to the unanticipated effectiveness of this second transnational campaign.

The Narmada Water Disputes Tribunal, examined in the previous chapter, did not explicitly address environmental issues in its proceedings, but many of the decisions it made were "triggering mechanisms" for almost all subsequent environmental controversies over the Narmada Projects.[1] It was during the ten years of the Tribunal's hearings, moreover, that development dynamics had begun to dramatically change as a result of emergent transnational norms, if not practices, with respect to the environment, human rights, and indigenous peoples. A decade earlier, the view that these issues were central to development, particularly with respect to the environment, was either extremely marginal or did not exist at all. By the early 1980s, however, procedural and organizational structures incorporating these issues had been established in the World Bank and numerous countries around the world, including India.

A massive growth in nongovernmental organizations and social movements working domestically, internationally, and increasingly transnationally, accompanied these changing norms (see chapter 1). This growth directly contributed to the formation of the transnationally allied opposition to the Narmada Projects that emerged in the mid-1980s. While building on the first transnational reform campaign examined in the previous chapter, this second coalition was primarily dedicated to halting construction on the projects all together in favor of alternatives based on a completely different vision of development.[2] But ultimately, it was grassroots mobilization in the context of India's battered but resilient democratic institutions that made the crucial difference in struggles over big dam building and development in the Narmada River Valley.

TRANSNATIONAL ENVIRONMENTALISM IN INDIA AND AT THE WORLD BANK

The United Nations Conference on Environment held in Stockholm in 1972 put environmental concerns squarely on the international agenda, even though a sharp division over the orientation of the new agenda between first and third world countries developed.[3] In India the process of integrating environmental factors into development policy had begun even earlier with "The Fourth Five Year Plan" (1969–1974), which was completed in 1970. In the plan, Indian authorities formally stated for the first time that "planning for harmonious development . . . is possible only on the basis of a comprehensive appraisal of environmental issues," although no concrete changes in actual development practices were proposed consistent with this ecological principle.[4]

It was only after the additional legitimation provided by the Stockholm Conference, however, that the federal Government of India actually translated these novel norms and principles into procedural and organizational structures. It first established the National Committee on Environmental Planning and Coordination (NCEPC) within the prestigious Department of Science and Technology in 1974 to serve as an advisory body on environmental issues facing the country. Chaired by Prime Minster Indira Gandhi herself, the NCEPC established various task forces to deal with select issues, such as identifying environmental problems, carrying out environmental investigations, and integrating general environmental concerns into specific development programs and projects. At this point, few Indian officials were in favor of devoting state resources to environmental issues, and self-conscious ecologically minded mobilization of the civil society was just emerging.

Thus, the formal institutional changes within the Indian state with respect to the environment were substantially motivated by progressively spreading global principles.[5] Additional environmental institutions followed quickly. India's Parliament drafted "The Prevention of Air Pollution Act" in 1972, and in 1974 passed "The Water (Prevention and Control of Pollution) Act." The federal government also quickly realized that most constitutional authority related to environmental issues resided with the States. Thus, in 1976, it transferred forests from the State List to the Concurrent List (shared federal-state authority). In addition, by 1977, based on recommendations from the NCEPC, Prime Minister Indira Gandhi issued a directive that environmental impact assessments be completed by federal agencies for all medium and major irrigation projects—including all big dam projects. This trend culminated in 1980 when, based on the recommendations of the high level federal government appointed committee on India's environment, the Prime Minister created a federal-level Department of Environment, and Parliament, passed the Forest Conservation Act.[6]

The trajectory of the Narmada Projects would not have been significantly altered had these procedural and structural changes in the Indian state not occurred. All the components of the scheme, including the Sardar Sarovar Project in Gujarat and the Narmada Sagar Project in Madhya Pradesh, now had to acquire environmental and forest clearances—the expanded version of the environmental impact assessment that had been mandated in 1977—from the Department of Environment. These were added to the techno-economic clearances from the Central Water Power Commission, investment clearances from the Planning Commission, and final clearances from the Prime Minister's Cabinet that had historically been required. The 1980 Forest Conservation Act also imposed restrictions on diverting (reserved) forest areas for other uses, such as reservoirs created by big dams. Before its enactment, Gujarat had established two resettlement village sites for Sardar Sarovar oustees by clearing forest areas, a practice that would have continued unhindered had the Forest Conservation Act not come into effect and the Department of Environment not been created to monitor and implement it.

The World Bank also incorporated a number of environmental procedures and structures into its organizational framework, at least on paper, at the same time it became involved with the Sardar Sarovar-Project. In preparation for the Stockholm Conference, then President Robert McNamara appointed an environmental advisor to create an environment and health unit at the Bank, primarily to conduct research on these issues. The following year an Office of Environmental Affairs was established.[7] Based on the investigations and recommendations of this unit, the Bank subsequently amended its project appraisal procedures "to caution against the selection of projects that might have excessive social and environmental costs," and "Loan officers were given special responsibility to ensure that all issues related to socio-culturally relevant institutions and the protection of the environment were properly considered."[8]

By 1984, the World Bank had issued its most comprehensive ever policy of social and environmental guidelines with respect to the funding of development projects in the form of operational manual statement (OMS) 2.36. The new code called for the inclusion of ecological factors from the earliest stages of project formulation and prohibited the Bank from supporting projects that would result in irreversible environmental damage, infringe upon any international environmental agreement to which the recipient country had adopted, or displace people without satisfactory mitigation measures.[9] When the Bank was reorganized into four regional divisions in 1987, an environmental unit was mandated for each region. In the process, the Office of Environmental and Scientific Affairs was transformed into the Environmental Department, which comprised a central office at the Bank's headquarters in Washington, D.C., and the four regional units.

A World Bank—Nongovernmental Organizations consultative committee was also established to improve relations between Bank staff and voluntary groups. In fact, a former Director of the Environment Department stated that nongovernmental organizations "were undoubtedly instrumental in bringing about the changes that were initiated" in the Bank at this point.[10] The multilateral development banks campaign of the 1980s, in which transnationally allied opposition to the Narmada River Valley Projects played a pivotal role, was one of the central means by which nongovernmental organizations worldwide motivated reforms in World Bank procedures and organizational structures (see chapter 6). While actual practices at the Bank were much slower to change than formal procedures and structures, these reforms also increased the institutional openings available to pressure and monitor the Bank. These provided critical points of leverage for the transnational opposition to the Narmada Projects in their lobbying efforts vis-à-vis the Bank.

As a result of these international and India-specific institutional changes related to the global spread of novel norms, the Government of Gujarat (GOG) hired the Maharaja Sayajirao (M.S.) University of Baroda in early 1983 to coordinate a benchmark report on the environmental aspects of the Sardar Sarovar Project, both for the environmental clearance it now needed within India and to satisfy the World Bank's new requirements that environmental factors had been investigated. Based on this report and information from the master plan for development of the Narmada River Valley submitted by Gujarat authorities to the Tribunal during its proceedings, the Sardar Sarovar Project was referred to the Department of Environment of the federal Government of India for clearance approximately six months later. A preliminary "Environmental Impact Study" of the other major dam, the Narmada Sagar Project, that was to be located upstream in Madhya Pradesh, was also submitted by the Government of Madhya Pradesh's (GOMP) Environmental Planning and Coordination Organisation shortly thereafter. Thus, within a period of less than five years, environmental issues were no longer absent from any discussion over the Narmada Projects.

These globalizing norms had clearly begun to alter the institutionalized context in which the Narmada Projects would be executed in India and at the international level. But the early studies and reports that were mandated by the newly established procedures were crudely prepared and provided easy targets for criticism from skeptics. A team of Delhi University students from the Hindu Nature Club and Kalpavriksh, a Delhi based environmental nongovernmental organization, were the first independent actors to complete a study on the impacts of the Narmada Projects. Based primarily on a fifty-day research trip they conducted through the Narmada Valley in July and August 1983, they suggested that the M.S. University Report had numerous problems and that "much of the information in the study derives from outdated Government sources rather than from fresh empirical studies." They proceeded to identify several areas that needed further examination, such as geological impacts, the impact on flora and fauna, and the treatment of catchment area forests, among others.[11] Citing Thayer Scudder's 1983 World Bank pre-feasibility report on resettlement, and based on their own research, the authors further argued that, "it is a callous mistake to let officials from the Irrigation and Revenue Department (of the GOG) handle resettlement," because "without a total understanding of the cultural ethos and psychological make-up of the tribal and the peasant, rehabilitation is bound to be a failure," and, "given the massive scale of the resettlement programme . . . is all this feasible?"[12]

Not only was this the first independent, nongovernmental analysis that integrated both environmental and social issues in its assessment, but also the two-level critique that emerged from this report foreshadowed the advocacy tactics that the second transnational coalition would increasingly employ against the Narmada Valley Projects. The first criticism concluded that there were serious problems in the formulation and implementation of the projects but that these could possibly, but not likely, be improved with more detailed studies in a variety of areas, a much greater deployment of resources, as well as better management and oversight. The second, and more profound, criticism questioned whether the projects really represented development, if development was to be participatory, equitable, and environmentally sustainable, in the context of the Narmada River Valley and India more broadly? The basis was thus set for the progressive integration of social and environmental concerns in the form of a broader challenge to the meanings of development that the Narmada Projects represented: in other words, the vision of development that the projects symbolized and instantiated.

This critique of the Narmada Projects coincided with, and contributed to, the emergence of transnationally allied opposition to big dam building around the world (see chapter 6). Ashish Kothari of Kalpavriksh sent a letter to Edward Goldsmith, one of the editors of *The Ecologist,* a widely respected environmental journal published in Great Britain, with a copy of the Hindu Nature Club/Kalpavriksh report on the Narmada Projects.[13] An abridged version of this report was published in *The Ecologist* in 1984, and was also included in Goldsmith and Nicholas Hildyard's subsequently published book on the negative effects of big dam projects, along with numerous reports of similar experiences from all over the world.[14] This book, *The Social and Environmental Impacts of Large Dams,* was not only dedicated to delegitimizing big dam building but also to promoting alternative participatory, socially just, and environmentally sustainable uses of water resources development worldwide.

The report also contributed to the formation of the second transnational anti-Narmada coalition and was the first tool used to pressure Indian authorities via the Department of Environment. It was distributed to various nongovernmental organizations in India, including the Gujarat based Arch-Vahini and various foreign nongovernmental organizations, such as Oxfam and Survival International, who gradually proceeded to incorporate environmental criticisms into their lobbying efforts around the Narmada Projects vis-à-vis the World Bank. Perhaps more importantly, at least at this point in time, the report was sent to the federal Government of India's Department of Environment. The Environment Department had already begun reviewing the environmental clearance applications for the Sardar Sarovar and Narmada Sagar Major Dam Projects and took a keen interest in the criticisms leveled in the report.

Had the federal Department of Environment not been established in 1980, partly as a result of the global spread and gradual domestic institutionalization of environmental norms in India, this political opportunity structure would not have been available to the anti-dam critics. Based on their emergent guidelines, early reviews and the Kalpavriksh critique, the Department of Environment's initial response to the GOG and GOMP was that their environmental assessments were incomplete and unsatisfactory. They provided the GOG and GOMP each with a long list of items on the Sardar Sarovar and Narmada Sagar Projects that would require more in-depth analysis.[15] During the period between 1985 and 1987, what by then had become the Ministry of Environment and Forests (MOEF) persisted in its refusal to grant environmental clearances to either the Narmada Sagar or Sardar Sarovar Projects, cornerstones of the broader scheme to develop the Narmada Valley.

Big dam proponents within India used the World Bank's support for the projects as a tactic to persuade the MOEF to grant the necessary clearances.[16] The World Bank had already approved a $450 million loan and credit agreement for the Sardar Sarovar Project in 1985. But by 1986, according to T. N. Seshan (then GOI's Secretary of Environment), "the fact that the MOEF was so skeptical about the environmental soundness of the projects, and the environmental clearance had become the main bottleneck within the Indian state preventing implementation of the Narmada Projects, convinced many activists working with the oustees in the Valley to take up environmental issues more seriously and intensively focus one set of lobbying efforts on the MOEF."[17]

This growing realization of the centrality of environmental issues and the Ministry of Environment and Forests in the struggle against the projects was later echoed by activist Medha Patkar:

> when I first entered the Narmada valley and a few villages, the situation was that the project work on the Sardar Sarovar (SSP) was stalled. It had been suspended at the behest of the Ministry of Environment and Forests owing to the nonfulfillment of basic environmental conditions and the lack of completion of crucial studies and plans. . . . So we came to Delhi and went to the Ministry of Environment and Forests. T. N. Seshan happened to be there and we found out what the real status of the project was, and it came to his notice that the project had not even been sanctioned by the Ministry. . . . So, a major question in our minds was that if this was the reality, how could the World Bank sanction the project and sign the agreements? [. . .] it was very clear that the World Bank had knowledge of many of the risks. . . . So we had to raise these larger issues.[18]

In fact, Patkar had taken the lead in forming the Narmada Dharangrast Samiti among the Maharashtra village oustees by February of 1986. Over the next year, she, along with other activists who joined her and the villagers themselves, ratcheted up their information gathering and mobilization activities in Maharashtra.

The activists also began organizing the villagers of Madhya Pradesh and forging links with the other domestic nongovernmental organizations (such as the Rajpipla Social Service Group and Arch Vahini) that had been working in the Narmada Valley since the early 1980s. The continuing investigation and incorporation of environmental issues into the struggle against the projects, Patkar later recounted, "were soon firmly rooted in the notion that sustainable resource use and control by local peoples, and the prevention against the encroachment on those resources by outsiders, was fundamental to ensuring our vision of participatory, socially just and equitable development."[19]

TRANSNATIONAL LINKS TO THE MULTILATERAL
DEVELOPMENT BANKS CAMPAIGN

A similar process of normative and strategic integration was also under way at the international level during this period. A multilateral development banks (MDB) campaign, as Lori Udall (formerly of the Environmental Defense Fund and then at the International Rivers Network) later wrote, in which "partnerships and alliances had been forged among NGOs in the World Bank's member countries with a common vision of sustainable development that is decentralized, socially just, and environmentally sound" had been ongoing from the early 1980s.[20] This campaign involved lobbying efforts to reform the projects, policies, and procedures of the World Bank, as well as other multilateral development banks, and it pressured these international organizations to become more democratic—i.e., more transparent about their decisions, participatory in their activities, and more publicly accountable. Moreover, Udall later stated that, "since 1985 when the World Bank approved the $450 million for the Sardar Sarovar Project, Narmada had come to symbolize destructive development," because "central to non-governmental organizations' criticisms of the World Bank is the Bank's lack of accountability to people directly affected by its projects and its promotion of large-scale, centralized, and outdated development projects that often are not fully completed or do not pay for themselves. . . ."[21]

The transnational anti-Narmada campaign quickly became a central vehicle for the anti-MDB coalition. While commencing her domestic mobilization efforts, Medha Patkar met Marcus Colchester of Survival International in the United Kingdom and Bruce Rich of the then Environmental Defense Fund in Washington, D.C., when they visited the Narmada River Valley on a trip to India in 1986.[22] Survival International had already linked its lobbying activities around the Narmada Projects with the first transnational reform coalition that had emerged in 1984–85. The Environmental Defense Fund had taken a lead in the multilateral banks campaign, and thus the Narmada Projects were of particular interest to that organization. As a result, Rich, who earlier had become aware of the problems with big dams through his participation in the 1982 Washington, D.C., "Dam Fighters Conference"—organized by U.S.-based environmental nongovernmental organizations, such as the Environmental Policy Institute and the American Rivers Conservation Council—invited Patkar to the United States to consolidate the transnational anti-Narmada campaign.

The sustained advocacy vis-à-vis the Narmada Projects by this expanding transnational coalition and linked set of campaigns produced increasingly dramatic results over the next five years. Patkar made her first journey to the United States in 1987. It was during this trip that plans for the Narmada International Action Committee—a more formalized transnational coalition among international and domestic nongovernmental organizations from around the world targeted at the Bank's Board of Executive Directors, other foreign donors, as well as the federal and State governments and bureaucracies in India—were formulated.[23] During her stay in the United States, Patkar met with World Bank senior officials in Washington, D.C., for the first time to discuss the severe problems with the Narmada Projects. As she recalled:

> The questions raised were not merely related to rehabilitation. One question raised was, for example, how could the World Bank sign the agreement when our own indigenous agencies, the Ministry of Environment and Forests and the Planning Commission, had not cleared the project? This was a distorted decision-making process and the influence of the World Bank then would make the project a fait accompli, and the sanction would be drawn out and the process was going on anyway. The World Bank had no answers. The only answer came from consulting the Indian legal advisor they had at the time, Mohan Gopal, who said, "Oh, we were not aware of this kind of a procedural requirement from the Ministry of Environment!"[24]

While the initial period in the transnational campaign focused on presentation of problems with the projects to the World Bank's operations staff, gradually a more systematic lobbying effort targeted at the Bank's Board of Executive Directors (and other foreign donors, such as Japan and the United States) was organized. This tactic of ratcheting up the pressure on Bank officials and staff in Washington, D.C., and holding them accountable to policies and procedures that had already been adopted, was vigorously applied by representatives of transnationally allied advocacy groups after Patkar returned to the struggle on the ground in India.

DOMESTIC PROPONENTS PUSH THE PROJECTS FORWARD

Simultaneously, in the first half of 1987 project authorities and State government officials in India had also begun to exert their own pressure, through the

Prime Minister's office and Ministry of Water Resources (MOWR), upon the Ministry of Environment and Forests (MOEF) so that it would issue the environmental sanctions for the Sardar Sarovar and Narmada Sagar Projects. As Dr. S. Maudgal, senior advisor to the MOEF recounted, "we received a lot of political pressure from the GOG and GOM through Prime Minister Rajiv Gandhi and the Secretary of Water Resources, Ramaswamy Iyer, to grant a conditional environmental clearance. The Chief Ministers, project authorities, and other proponents from the States would say, 'yes you are right about the necessity of environmental plans, but can't you give us the go-ahead and we will do whatever you want pari passu with (at the same time as) the construction of the project?'"[25]

A series of letters subsequently passed between the MOWR and the MOEF in which the former also pushed for a conditional clearance on the basis that the projects had already been given technical approval by the Central Water Commission. The Secretary of Water Resources and the Chief of the Central Water Commission forcefully argued that the projects had also already been sanctioned and funded by the World Bank, and how would it look if the projects were stalled by the Government of India itself?[26] Yet the MOEF, in particular its Minister Baganlal, Secretary Seshan, and Senior Officer Maudgal, stood its ground and sent a note to Prime Minister Rajiv Gandhi in April of 1987 recommending that the Sardar Sarovar and Narmada Sagar Projects not be given environmental clearance.[27] They noted the continued lacunae in environmental studies, planning and mitigation measures as clear evidence that the projects should not be cleared for implementation.

The Chief Ministers of the Gujarat, Maharashtra, and Madhya Pradesh Governments responded by arguing to the Prime Minister that his Congress Party would lose the next elections in these three states if he did not get the project approved. As Chimanbhai Patel, then Chief Minister of Gujarat later told me, "There was no way that it would not be sanctioned. We were committed to Sardar Sarovar Project—our lives depended on it. We knew Rajiv was weak, and we were ready to exploit his weakness."[28] Already reeling from the losses the Congress Party suffered in mid-term elections in a number of States, dependent on the leaders of these powerful States in western India and strongly supported by the Ministry of Water Resources, Prime Minister Rajiv Gandhi pushed the conditional environmental clearance through in June of 1987, and a forest clearance was accorded in September of the same year.[29]

The conditions placed on the environmental and forest clearances were quite expansive, at least on paper. More detailed planning and coordination based on further in-depth studies were required in the following areas: (1) rehabilitation, (2) catchment area treatment, (3) compensatory afforestation, (4) command area development, (5) surveys of flora and fauna, (6) carrying capacity of the surrounding area, (7) seismicity, and (8) health aspects. This was the first time in India's history that such wide-ranging conditions had been imposed on big dam projects. These studies and plans were to be completed by 1989 at the latest; and if not completed by that time, the clearance would automatically be revoked. Furthermore, the terms of reference for the Narmada Control Authority, which was mandated by the Narmada Water Disputes Tribunal's award in 1979 to oversee the projects, were expanded to "ensure that environmental safeguard measures are planned and implemented pari passu with progress of work on projects."[30]

Although the structure of political power relations within India's federal bureaucratic democracy facilitated the continuation of the Narmada Projects at the time, the gradually changing international and Indian development context interacting with the lobbying activities by transnationally allied nongovernmental organizations and peoples' groups was beginning to have a marked effect. Based on the augmented terms of reference, two subcommittees of the Narmada Control Authority were formed—one on the environment with the Secretary of the MOEF as chairperson and one on resettlement with the Secretary of the Ministry of Welfare as head—to monitor progress in these areas. Representatives from domestic nongovernmental organizations were mandated to be members of the sub-committees to ensure greater accountability to the people living in the Narmada Valley who would be affected by the projects. These newly incorporated procedures and structures created additional advocacy opportunities for the progressively more organized and mobilized transnational linked opposition coalition.

THE SPLIT BETWEEN THE TWO NARMADA CAMPAIGNS

By this time, the GOG had also begun to formulate the new resettlement resolutions in response to the intensive pressure of the first transnational reform coalition (see chapter 3). But this victory was of short duration. On December 27, 1987, Amarsinh Chaudary (by then Chief Minister of Gujarat) announced the reformed resettlement policy at a specially convened meeting of the Gujarat oustees at Kevadia Colony, the on-site headquarters of the Sardar Sarovar Project. Yet soon thereafter, the Deputy and Additional Collectors in charge of rehabilitation in Gujarat told the oustees that they were prepared only to give land to land-holding and not to all families, as per the previous November 1, 1985, Government Resolution.

The groups working and living in the Narmada Valley sent correspondence to and held meetings with officers of the Rehabilitation Agency of Gujarat's Narmada Development Department but with no substantive reaction from these higher level authorities on the about face in GOG policy. The GOG came out with a glossy brochure given to all oustees describing their rights and benefits under the revised resettlement plan. However, the last paragraph stated that claims by oustees would not be entertained on the basis of the contents of the brochure. Encouraged by the recent environmental clearance afforded the projects by the federal government, State officials and project authorities in Gujarat also initiated a massive public relations campaign with the distribution of five booklets promoting the Sardar Sarovar Project: (1) "SSP: The Lifeline of Gujarat"; (2) "SSP and the Environment"; (3) "SSP: A Ray of Hope"; (4) "SSP: A Boon to Gujarat"; and (5) "SSP: A Planned Ecological Harmony Amongst Man, Water, Land and Vegetation."

A widening rift quickly emerged between those who took a "no implementation of resettlement, no dam" stand and those who took a more unequivocal anti-dam position. The initial split occurred at a New Delhi conference organized by the World Wildlife Fund for Nature—India held in November 1987. The objective of the conference was to facilitate the coordination of activities among the various nongovernmental groups and organizations contesting the Narmada Projects inside and outside of the Valley. While some groups, such as Arch-Vahini and Anand Niketan Ashram, subsequently began to cooperate with project authorities on implementation of the reformed policies on resettlement, others such as the Narmada Dharangrast Samiti and Kalpavriksh took a stand that a complete and comprehensive reevaluation of the Narmada Projects was required before any further construction should occur. The latter groups combined a number of critiques: that the resettlement reforms would not be implemented, that the projects would cause irreparable environmental damage, and that the projects' costs far outweighed their expected benefits.

This second transnational anti-Narmada coalition thus launched an even more massive multilevel, transnational campaign against the projects as symbolic of a path of destructive development that had to be stopped, no matter what the costs.[31] Patkar and other activists began mobilizing the villagers of Madhya Pradesh even more intensively by holding progressively better attended and organized public meetings and protests in that State. This was no easy task, because while the people to be displaced in Gujarat and Maharashtra were primarily tribals/adivasis, the population in the Narmada River Valley in Madhya Pradesh was an extremely complex mix of rich "green revolution" farmers, peasants, and various tribal/adivasi communities speaking a range of different languages, with heterogeneous cultural traditions, and high levels of economic inequality and degrees of social stratification.[32]

Yet gradually, village by village, the sense of common doom if the Narmada Projects were completed, combined with the inspirational and tireless commitment of the activists, convinced the villagers to work together. The common views that emerged among the villagers were that the benefits of the Sardar Sarovar Major Dam (and other such large development projects) were appropriated by rich and powerful groups while the poor, lower castes, and tribals bore the brunt of the costs; that it was puzzling that the politicians and bureaucrats asked them to sacrifice for the progress of the country, when they had largely been prevented from participating as full citizens in India's political system, economy, and society for so long; and that power generation was not of interest to them because they could not eat electricity.[33] This is not to suggest that disagreements did not continue among the various groups that constituted the movement in the valley, but rather that through the process of struggle, the foundation of a collective set of understandings began to emerge.

In addition to organizing in the villages of Maharashtra and Madhya Pradesh, activists began mobilizing the people who would be displaced by the construction of the canal system for the drought-prone Saurashtra and Kaachch regions of Gujarat. This tactic also led to the heightening skepticism by people in these regions that promises by Gujarat State officials of water being delivered to them from the Sardar Sarovar Project were gross exaggerations. These efforts culminated in a major rally on January 30, 1988, at the dam construction site of Kevadia colony where it was decided by the various groups, who were to be displaced, that none of the families would voluntarily resettle unless all the questions about the economic, social and environmental aspects of the projects were reviewed and answered by domestic authorities.

At the same time, critics continued with their own independent investigations of the projects. One of the major sources of information to the domestic opposition was the World Bank reports that foreign nongovernmental organizations like the Environmental Defense Fund and the World Bank Information Center acquired and passed on to their domestic counterparts. Without this crucial source of information, the movement in the Narmada Valley and throughout India would have been considerably weakened, because officials from the Narmada Control Authority often denied them access to the main reports of domestic experts and committees. An additional source of information that also proved crucial was made up of the nongovernmental members on the Environment and Resettlement Sub-Committees of the Narmada Control Authority whose participation had been required by the 1987 conditional environmental clearance. Gradually, a large archive of documents and studies were gathered, materials that were to be used effectively to criticize the technical, economic, social, and environmental aspects of the Narmada Projects.[34]

The opposition grew stronger with each passing day. A meeting of leaders and representatives from all the villages, which also included the activists, was held on April 24, 1988; it discussed the probable non-implementation of the resettlement program. A "Patrika Rally" was promoted in which the local residents, over the following days and weeks, sent back the various promotional brochures about the project that the authorities had passed out to them. Subsequently, a convention was held on May 14, 1988, at which 1,500 people, including representatives from Maharashtra and Madhya Pradesh, gathered to protest against the failure of the GOG to implement its own 1987 resettlement reforms. About 30 leaders from different villages spoke vividly and precisely about the problems they were facing. After a unanimous vote, it was decided once again that if rehabilitation and resettlement of oustees of Gujarat were sought to be downgraded by the State governments, then those to be displaced in Madhya Pradesh and Maharashtra would fight "shoulder to shoulder with their brothers and sisters."

On the morning of May 15th, practically all the newspapers in Gujarat in Gujarati, Hindi, and English carried the message of the Convention, "no dam, if no rehabilitation," on their front pages.[35] Medha Patkar recounted the building anti-dam sentiment:

We then decided, after a six-hour marathon discussion in 1988, that we must give two months time to the government and, then if we did not get answers, oppose the dam. On the same day in the evening that we had the long discussion with the Narmada Control Authority and the three states' and Centre's secretaries, I vividly remember that the Gujarat organizations, especially Arch-Vahini, had also come. Rather, their activists were brought there by some of the leaders of the Gujarat villages, who were very keen that there be unity amongst all of us. They knew us since we used to walk through their villages and stay in their houses. These leaders said, "Agar ye nahi hua to, Diwali tak nahi hua to hum Dam ko rok denge" (If the government does not give answers till the Diwali festival in October we will stop this dam). After that meeting, we signed a common letter saying that we were giving the government two months time, and if they did not answer all the questions in two months, we would oppose the project. This was handed over by the Vahini activists to the secretaries from the center sitting in the circuit house on the 14th or 15th of May 1988.[36]

On May 15, representatives of the Narmada Control Authority (NCA) visited Kevadia Colony, the location where the convention had strategically been held the day before. After being forced to listen to the complaints of the local representatives, the Chairman stated that the December 1987 Gujarat Government resolutions were beyond the jurisdiction of the NCA and it could not therefore ensure their full implementation.

The grassroots leaders and representatives from the three States sent a joint letter to the NCA and the various State project authorities reiterating their refusal to be treated separately in terms of resettlement any longer. On May 19th, the Gujarat newspapers carried a story in which the Gujarat Chief Minister assured that the resettlement policy based on the December 1987 resolutions would be implemented and denounced the actions of the lower level bureaucrats.[37] While the Gujarat based organizations were satisfied with this response of the authorities, with Arch-Vahini, in particular, shifting to a policy of assisting the GOG with implementation of its revised and more expansive resettlement reforms, the organizations and groups led by Medha Patkar felt that the resolutions did not amount to nearly enough.[38] Though many of the villagers were persistent in their efforts to keep the groups united, in the end the differences were too deep to overcome.

THE NARMADA BACHAO ANDOLAN (SAVE THE NARMADA MOVEMENT)

The grassroots anti-dam movement that coalesced to halt the Narmada Projects was even more broad-based and relied even more heavily on mobilization by the residents of the Narmada Valley than previous episodes of domestic resistance. From the summer of 1988 onward, the domestic struggle against the projects moved into full swing. On August 18, 1988, full opposition to the projects was announced at six cites across the three Indian states of Gujarat, Maharashtra and Madhya Pradesh. The tactics chosen domestically were Gandhian nonviolent ones, including more comprehensive organization of the people to be negatively affected by the projects, noncooperation with project authorities, the continued investigation of flaws with the projects, and the formulation of alternative development initiatives which would be more participatory, socially just, and environmentally sustainable.

In September, demonstrating the strength of the coalition of domestic nongovernmental groups all across India supporting the struggle against the projects, major protests and actions were launched throughout the country. Despite this visible increase in mass mobilization inside and outside the Narmada Valley domestically, the Government of India's Planning Commission granted an investment clearance of Rs 6406 crore for the projects on October 5, 1988, of which only Rs 265 crore, or less than 3 percent, was allocated for resettlement and even less for environmental studies and mitigation efforts.[39] In order to quell the spreading agitation, the GOG also imposed the Official Secrets Act on 12 villages in the Narmada River Valley, ushering in a reign of repression that increased in severity over the coming years. The act gave authorities wide-ranging powers to control access to information, prohibit public protest, and even imprison opponents on the grounds of threat to national security.[40] In response, a petition that questioned the act's constitutionality was filed by activists in the Gujarat High Court, an avenue that was available, given India's still relatively resilient democratic institutional framework.

The domestic mobilization against the Sardar Sarovar Project only grew in size and strength. Sitarambhai, an adivasi leader of the struggle, claimed that "the two responses of the government were first to try to co-opt through money/positions and second to repress/crush us. Both were tried, both failed."[41] In the middle of November 1988, over seventy activists, journalists, and scholars undertook a Sanwaad Yatra or Conscientization March through fifteen of the villages, holding public meetings along the way. This yatra was followed up by a flurry of activity including workshops, rallies, street plays, letter writing, and press reports to increase public awareness throughout India about the projects, increasingly organized and implemented by the local people themselves. On January 30, 1989, 500 people from the Narmada Valley protesting the Official Secrets Act were arrested by Gujarat police and, that same day, dam site workers went on strike against the human rights abuses perpetrated by the contractors constructing the Sardar Sarovar Project.

The combination of mobilization by the people of the Valley and project workers, publication by nongovernmental organizations as well as the Gujarat High Court of fact finding reports revealing human rights abuses, and the continued pressure on Indian authorities from transnationally allied nongovernmental organizations outside India via the World Bank, compelled the GOG to grudgingly withdraw the Official Secrets Act on March 29, 1989. As Vijay Paranjpye, an economist who conducted a benefit-cost analysis of the Sardar Sarovar Project for a domestic nongovernmental organization (INTACH) reported:

> In a desperate attempt to stem the agitation, the government resorted to violence and assaulted eight activists. They also arrested 4,000 tribals who were among the 6,000 persons who converged upon Waghadia after a three-day long march. This triggered off a dharna in Bombay, a relay hunger strike in Madhya Pradesh and meetings of intellectuals and concerned citizens to express their solidarity with the oustees' cause throughout India.[42]

Consequently, domestic big dam opponents won an important victory by restoring basic democratic rights in India.

The "Narmada Bachao Andolan" or "Save the Narmada Movement" was formally constituted during the following spring and summer. It united most, though not all, of the organizations and groups that had been working in the Narmada Valley since the early 1980s. The whole series of events culminated in the Harsud rally of September 28, 1989. As activist, Sripad Dharmadikary, recounted; "There was a huge rally organized in Harsud where a convention against destructive development, not just the Narmada, was drafted and signed. More than 300 nongovernmental organizations and 60,000 people came from all over the country, perhaps the largest rally on this issue ever held in India."[43] The "sankap" or resolve that was coined at the event was "Vikas chahiye, vinash nahin!" or "we want development, not destruction." Sympathetic observers identified the rally as the "coming of age of the Indian environmental movement."[44]

Indeed, it was this grassroots mobilization by the people that ultimately endowed the domestic anti-dam movement and the transnational anti-Narmada coalition with the legitimacy and power to stall the Sardar Sarovar Project, but not for a number of years to come. Activist Himanshu Thakkar described the evolution of the Save the Narmada Movement during this period in vivid terms:

> There were some exceptional minds and hearts working full time at it, people whose commitment and dedication was total and complete, who attracted diverse supporters from various places to do their own little part in whatever way they could. No institutional support was accepted, no foreign funding was taken in order to remain independent and free from castigation as being anti-nationalist. But the villagers themselves were the movement's true strength. Within the four years between 1985 and 1989, a multifaceted campaign was developed working at so many levels of governance—village, tehsil, state, federal even global . . . [45]

THE CONSOLIDATION OF THE TRANSNATIONAL ANTI-NARMADA COALITION

The campaign in the international arena had simultaneously grown. The Environmental Defense Fund, in a submission to a U.S. Senate Subcommittee in June of 1988, declared that the people in the Narmada River Valley were opposed to the Narmada Projects and that the World Bank did not have the right to proceed with construction without a fresh reappraisal. By autumn of 1988, convinced that Bank management and staff were not serious about reviewing or reforming the projects, Washington, D.C.-based campaigners took their case directly to the Board of Executive Directors and the Bank's Senior Vice President for Operations, Moeen Qureshi. Qureshi responded by leading a high-level Bank resettlement mission to India. Prior to his departure, the allied Washington-based nongovernmental organizations briefed Qureshi about the grassroots situation from the information sent to them by their domestic Indian counterparts—information that the Bank itself did not have—and persuaded him to meet with the affected peoples and their supporters in the Narmada River Valley.

During his visit to New Delhi, Qureshi met with Indian government officials, project authorities, as well as with Medha Patkar and other representatives from the Save the Narmada movement. After returning and holding several more meetings with Bank management, government officials, and nongovernmental organizations, Qureshi was convinced. He sent a letter on November 28, 1988, to Chief Minister Arjun Singh of Madhya Pradesh. In the letter, Qureshi threatened to suspend Bank support for the projects by March 1989 unless a comprehensive plan including "prompt identification of suitable lands for resettlement, careful preparation of rehabilitation options for landless people and designs for development of relocation sites," was formulated.[46]

The Bank was clearly concerned about the persistent problems with and growing conflicts over the projects. In April of 1989, Thayer Scudder was once again called upon by the Bank to accompany a mission led by then Chief of Agricultural Operations in the India Department, Jan Winand. In his memorandum, Scudder remarked that, "In comparing the August situation with the situation today, I believe that there has been a very serious deterioration," and further stated, "I doubt that the three state governments can carry out an R & R Program that will meet minimal World Bank Requirements." He thus recommended at least temporary, if not permanent, termination of loan disbursements.[47] "However," as Lori Udall of the then Environmental Defense Fund recounts, "Scudder's recommendations were ignored. . . . Executive directors had no access to his recommendations and only heard of them a year later when Scudder met with them and recommended a suspension of the project."[48]

The Washington-based nongovernmental organizations also pursued another line of pressure via the United States House of Representatives Subcommittee on Natural Resources, Agricultural Research, and Environment. Chairman James Scheuer, Democratic Party Congressman from New York, organized a special oversight hearing to investigate the World Bank's support for the Narmada Projects. On October 24, 1989, testimonies were delivered by three Indian activists: (1) Medha Patkar of the Narmada Bachao Andolan; (2) Girish Patel, a human rights lawyer and head of the Gujarat nongovernmental organization, Lok Adikar Sangh; and (3) Vijay Paranjpye, the economist who had been commissioned by the INTACH to conduct an independent benefit-cost analysis of the projects. In addition, Frank Vukmanic, head of the Office of Multilateral Development Banks in the U.S. Treasury, as well as Peter Miller and Lori Udall of the Environmental Defense Fund based in Washington, D.C., spoke at the hearings. While the World Bank as an international organization was not legally bound to appear before the legislature of any country, even Scheuer's offer to the Bank to testify off the record was refused.[49]

Issues including the proposed costs and benefits of the projects, the impacts on the tribal peoples and others to be displaced, the negative environmental consequences, and noncompliance with Bank and Indian agreements and operating procedures were discussed. Medha Patkar's dramatic testimony on these issues "was applauded by a packed room of congressional staff, journalists, environmentalists, and human rights activists."[50] Intervening late in the hearing, Chairman Scheuer summed up his views:

> Well, it seems to me that when you have a loan and a project that encompasses literally generations, and it's been approved on a certain set of criteria and cost-benefit analyses are provided, and assurances of environmental nondestruction are provided, and assurances of humane treatment of native populations are provided, and when after the fact, after perhaps a 10-year period—perhaps after a 10-year period, it's shown that these assurances and these protections and these precise measurement aids or cost-benefit analyses are in error in making financial judgements, if they are not real, and if those commitments to provide information . . . if they are not provided, it seems to me after a certain period of time the Bank has a right, and indeed the obligation, to say, "Hey. We're going to have a painful reconsideration of this whole project. It has gone amuck."[51]

Based on the hearings, a number of members of Congress sent letters to then President Barbar Connable of the World Bank calling for a suspension or outright cancellation of funding for the projects.

The Congressional hearings and Scheuer's interventions also began to shift the perceptions of other foreign donors. Congressman Scheuer himself went a step further and raised the issue at a Washington, D.C., meeting of Global Legislators for a Balanced Environment (GLOBE), a transgovernmental coalition, and subsequently sent letters to several European and Japanese funders to take the case up in their own countries. These included the OECF of Japan which was to fund turbines manufactured by the Sumitomo and Hitachi corporations, KFW of Germany which was involved in fisheries development, CIDA of Canada which had been approached on the issue of the environmental impact assessment, and ODA of the United Kingdom, which was called upon to conduct downstream impact studies on the projects.

The transnationally allied lobbying efforts were directed not only at the U.S. Government but also at the Board of Executive Directors of the World Bank. Soon after the Congressional hearings, the Dutch Executive Director to the World Bank, Paul Arlman, began to hold informal gatherings with nongovernmental organizations about the Narmada Projects. In October 1989, Arlman hosted a meeting between the Bank's Executive Directors and Medha Patkar where she presented the ground realities with respect to the projects. After hearing Patkar out, one Executive Director said, "When I hear what NGOs say about this project and then what Operations staff say, it sounds as if they are talking about two different projects."[52] Patkar and Arlman started a regular correspondence to ensure that the Board had access to other sources of information and this, along with the thousands of letters being sent to the Executive Directors from concerned individuals and nongovernmental organizations from their own countries, convinced many of them to monitor the projects more closely.

Encouraged by the events in Washington, and after traveling to India to examine the ground reality in the Narmada River Valley for themselves, Japanese nongovernmental organizations led by Friends of Earth (Japan), which had been a member of the Narmada International Action Committee, hosted an International Narmada Symposium in April 1990. Bringing together over 500 people, including activists from India, Japan, and around the world with Japanese Diet members, lawyers, and academics, the symposium received widespread coverage from the Japanese press and captured the attention of the Japanese public.[53] Following the symposium, representatives of the Narmada International Action Committee and other Japanese nongovernmental organizations held a series of meetings with officials from Japan's Ministries of Foreign Affairs, Finance, International Trade and Industry, and the Overseas Economic Cooperation Fund. The OECF had already lent millions of dollars for the Sardar Sarovar Projects and was in the process of considering more funding.

But three weeks after the symposium and follow-up lobbying efforts, the Japanese Ministry of Foreign Affairs announced that it would not provide further assistance to the projects because of domestic Indian opposition, lack of resettlement planning, and the absence of a comprehensive environmental assessment. The Japanese nongovernmental organizations did not stop with this victory but continued to pressure Japanese officials to take a more vocal stand on World Bank lending to the projects as well. As a result, on June 26, 1990, over twenty Japanese Diet members wrote a letter to President Barbar Connable arguing for suspension of Bank funding.[54]

Indian authorities were now being pressured from seemingly all angles. For example, the letter sent by the Japanese Diet members to the World Bank—which was motivated by the activities of the Narmada International Action Committee—was forwarded to M. S. Reddy, the then GOI Secretary in the Ministry of Water Resources, by Heinz Virgin, Director of the Bank's India Department, with the following message:

This letter lists important issues which are causing concern among members of the Japanese National Diet. . . . As you know, India has repeatedly assured the Bank, and recently the Executive Directors of the Bank, of the completion of a satisfactory and comprehensive R&R plan. . . . We would like to receive further assurances that linkages between dam construction and the R&R are still carefully synchronized and implemented. . . . Without such assurances, support from the World Bank will be difficult to sustain.[55]

Yet, Indian authorities remained steadfast in their commitment to the Narmada Projects and forged ahead with construction, despite the warnings of the World Bank, the lobbying of nongovernmental organizations all over the world, and the continued widespread domestic resistance. The reasons for this intransigence included the range of powerful groups in India that had a stake in the projects, the disbelief that the Bank would ever withdraw its funding and support, as well as, at the deepest level, a profound conviction among most politicians, bureaucrats, and much of the Indian population that big dams were synonymous with development. Each of these justifications was being challenged by the domestic movement on the ground.

STALEMATE IN THE DOMESTIC ARENA

The domestic mobilization in India had entered an even more hectic phase after the phenomenally successful Harsud Rally. Activists, villagers, and local people increasingly pressured various governmental agencies at the federal level through the MOEF and MOWR, as well as at the State level through the Narmada Control Authority—especially the Environmental and Resettlement Sub-Groups—to halt construction on the Sardar Sarovar Dam. The movement also continued to organize rallies at the village level, investigate the projects and alternatives to it, and inform the wider Indian public through the press, newsletters, and other media. On March 6 and 7, 1990, for example, ten thousand people blocked the Bombay-Agra highway that went through the Narmada River Valley for 28 hours. During the blockade, a memorandum demanding a comprehensive review of the projects was drafted and sent to the Chief Ministers of the four involved States and the press. In response, the then Chief Minister Patwa of Madhya Pradesh promised to have an independent review of the Projects—the first affirmative response to the long standing demand for a comprehensive reappraisal.

The mobilization activities at the State level did not abate. Between March 29 and April 5, Medha Patkar and a number of other activists participated in a fast in Bombay to protest against the failure of Maharashtra project authorities to formulate detailed resettlement plans for the people to be displaced in that State. The fast was called off when Chief Minister Sharad Pawar of Maharashtra sent a written letter promising that no villagers would be displaced until alternative lands were made available. But as with many others before and after them, neither of these promises was kept.

Failing to effectively alter the practices of State authorities, the Narmada Bachao Andolan decided in May 1990 to go straight to the leader of India's federal Government in New Delhi with their grievances and demands. V. P. Singh had just become the Indian Prime Minister of a center-left coalition government at the federal level on a social justice platform in 1989. As a result, many leaders of the movement believed that they would get a more sympathetic response from him and his administration. Hundreds of protesters congregated in front of the Prime Minister's residence in New Delhi and following a four-day sit-in, V. P. Singh relented and invited all the protestors into his residence. After hearing testimony from tribal leaders, representatives from the rich-farmer areas of Nimar, Medha Patkar, and Baba Amte, one of India's most renowned Gandhians who had recently joined the ever-growing movement, the Prime Minister agreed to initiate a comprehensive review of the projects.[56]

The Chief Minister of Gujarat State, Chimanbhai Patel, responded immediately, declaring that he would not tolerate the federal government's interference and organized a counter rally of project supporters in New Delhi. As in 1986–87, when then Prime Minster Rajiv Gandhi was compelled to push through the environmental clearance for the projects, V. P. Singh's leadership of the minority coalition government in 1989 depended heavily on parliamentary seats from Gujarat. Realizing he could not win a no-confidence vote in Parliament without the support of Chief Minister Patel, the prime minister retracted his commitment for a comprehensive review. As Patel later argued, "Gujarat and Maharashtra are two of the most industrialized and politically powerful States in India; no one can dictate what we do from the Centre."[57] Instead, by the end of the monsoon in 1990 after the waters began to recede, Singh asked only for more information to be made available to the public. The Andolan's efforts for a comprehensive review were once again blocked.

These barriers at various levels and agencies of India's polity were combined with widespread violation of human rights in Gujarat, Maharashtra, and Madhya Pradesh, and continued construction of the Sardar Sarovar Project.[58] Consequently as Amita Baviskar, a scholar-activist who worked in the Narmada Valley, noted at the time, "with the construction of the dam wall continuing unimpeded, the Andolan . . . had to race against time, organizing ever-larger, increasingly dramatic protests. . . . In December 1990, the Andolan set off from

Nimar on the Jan Vikas Sangarsh Yatra (March of the Struggle for People's Development) to peacefully stop work at the dam site."[59] More popularly known as "The Long March," thousands of local people, activists, and supporters walked for six days through Madhya Pradesh on their way to the Sardar Sarovar Dam site in Gujarat. But they were met at the border between the two States by a force of Gujarat officials, project authorities, military, and police.

Behind the police and military were thousands of Gujaratis who had been mobilized by government officials to conduct a pro-dam rally. The participants of "The Long March" were threatened with dire consequences if they crossed the border. In the words of one woman villager:

> There was a line drawn at the border that did not exist before. Gujarat asked the GOMP to arrest us but they refused. So when the march reached Gujarat, Chief Minister Chimanbhai Patel had mobilized forces to meet us at the border. They organized a huge rally on the border—had district collectors bring busloads of people. They were saying, "we will finish them off (the protesters) like a dobi (clothes washer) cleans off dirt—we will throw them into the Narmada." We stopped just a few steps before the border and a delegation was sent to ask, "why are we being harassed? This is a peaceful march and our democratic right as citizens of India." They responded, "no, you are going to act violently and destroy the dam." So we made camp on the Madhya Pradesh side of the border with little food and little resources while they made a huge, luxurious camp on the Gujarat side providing food, drink, even money from the State.[60]

Various delegations were sent by the marchers to attempt peaceful crossings of the border but all were turned back by the pro-dam representatives.

The blockage of the border provoked more drastic tactics; in protest, Medha Patkar and four others went on an indefinite fast. Officials responded by offering to do anything on R&R as long as construction was not delayed, but the Andolan no longer believed these types of promises. As days passed, there was increasing pressure from the federal Government of India and the World Bank on the State governments to negotiate with the protesters. The World Bank had received information about the indefinite fast via nongovernmental organizations based in Washington, D.C., that were in constant communication with domestic Indian nongovernmental organizations in Delhi that, in turn, were in touch with the movement in the Narmada Valley.

The political drama heightened in intensity. There was no forthcoming announcement of the Bank's plans to review the project, nor was there a positive response from Indian officials as the four fasters grew weaker. Washington-based nongovernmental organizations desperately attempted to convince World

Bank Vice President Moeen Qureshi to announce a reappraisal in the hope it would break the impasse on the border. The Andolan was privately assured that the World Bank would indeed constitute a review panel shortly thereafter. In the meantime, two of the fasters had been taken away and, on January 25th, a massive set of raids by Gujarat, Madhya Pradesh, and Maharashtra police occurred. Hundreds of protesters were beaten and arrested. The Madhya Pradesh police attempted to remove the remaining fasters but were blocked by a wall of protesters that formed around them. The Madhya Pradesh police threatened force but ultimately gave up.

Finally, as the health of the fasters continued to deteriorate, an all-India review panel for the projects was announced in Delhi. The protestors decided to call off the fast that had lasted a total of twenty-two days. Empowered by the announced World Bank and Government of India reviews, the Andolan withdrew on January 30th to the villages in the Narmada Valley proclaiming, "Hamare gaon me hamare raj" or "our rule in our villages." The slogan symbolized a challenge not to allow anyone into the villages without the Andolan's approval and a commitment that the local people would not move from their homes. During February and March of 1991, the GOMP unexpectedly notified villagers in the submergence zone that they did not belong to the State of Madhya Pradesh; this sparked a series of clashes with government census takers. These types of interactions between officials and the villagers, with attempts at coerced relocation being met by resistance, and resistance being responded to by repression, became a regular feature of life in the Narmada Valley by the early 1990s.

THE TRANSNATIONAL ANTI-NARMADA COALITION RAISES THE STAKES

In order to galvanize the campaign at the international level, a number of European nongovernmental organizations, led by Action for World Solidarity in Berlin, coordinated a tour for Sripad Dharmadikary and Kisan Mehta of the Narmada Bachao Andolan through Germany, Sweden, Finland, Denmark, the Netherlands, and the United Kingdom. Through meetings with parliament members and other government officials, various nongovernmental organizations and the press, the tour led to increased European criticism of the projects. As Sripad Dharmadikary later noted, "This was a critical period precisely because the structure of the World Bank's Independent Review of the projects was being shaped. We could not let any opportunity to lobby at the international level pass."[61]

Letters to the World Bank from members of the Swedish Parliament calling for a suspension of the projects, and a letter from the European Parliament pressuring the World Bank to fulfill the terms of reference for the Independent Review which had been promised during the Long March, were drafted and sent. After a short list of names was put forward by the Bank to chair the Independent Review, nongovernmental organizations around the world lobbied through their World Bank Executive Directors for nominees that were acceptable to the Andolan and Indian advocacy groups. In conjunction with the sympathetic involvement of a number of Executive Directors, particularly Eveline Herfkins of the Netherlands, the lobbying efforts eventually compelled the World Bank to announce on June 17, 1991, that an ex-UNDP Director Bradford Morse would head the Independent Review and, soon thereafter, that Canadian human rights lawyer Thomas Berger would take the position of deputy chairman.[62]

Since the Independent Review was the first ever conducted on an ongoing Bank project, it would likely set a precedent for such initiatives in the future.[63] The outcome of the Independent Review was also critical because it represented a potential institutional transformation at the World Bank and a major victory for the emergent transnational anti-dam network (see chapter 6). Consequently, in August 1991, representatives from nongovernmental organizations such as Lokayan—India, Friends of Earth—Japan, International Rivers Network—the United States, Probe International—Canada, and the Bank Information Center—Washington, D.C., met informally with the review team of Morse, Berger and Donald Gamble, an environmental policy analyst hired to assist with the investigation in Washington to present their perspectives on the Narmada Projects.

The likelihood of the Independent Review's potential impact was further increased by a number of decisions lobbied for by the transnational anti-Narmada coalition including: (1) the team was given a separate $1 million budget and an extended period of time to conduct its investigation; (2) the team was to have complete access to all information from the Bank and the various Indian governments; and (3) the results of the review were to be published independently of the World Bank. The transnational coalition also petitioned that the Review's terms of reference be broadened to include techno-economic viability questions but the Indian federal and State governments agreed only to a review confined to resettlement and environmental issues. Given the fact that environmental issues had not been even considered by the Narmada Water Disputes Tribunal during the 1970s or in its 1979 ruling, their matter-of-fact inclusion in the Independent Review demonstrates the remarkable change that had taken place in development dynamics during the intervening decade.

The domestic campaign in the Narmada Valley to resist evictions, however, had further intensified. In spring 1991, the GOG announced that rising waters during the summer monsoon would begin submerging the villages of Gadher and Vadgam in Gujarat. But the GOM did not give notices to the residents of the villages in Maharashtra who would also face submergence. The Andolan with the support of other domestic nongovernmental organizations petitioned the Bombay High Court to order the GOM to explain the premature and un-notified displacement process, while simultaneously organizing a series of rallies in all the villages to be submerged. The previous slogan of "koi nahi hatega, bandh nahi banega" or "no one will move, the dam will not be built" was replaced by the even more dramatic, "doobenge par hatenge nahin" or "we will drown but move we will not."[64]

The village of Manibeli became the central site for this new tactic of resistance, called jal samarpan, a protest of suicide by drowning, because of its proximity to the dam site and its being one of the first villages to face submergence. On August 3, Maharashtra police entered Manibeli and arrested 79 people under sections 144 and 151 of the Indian Penal Code. A team of Andolan members went underground vowing that "Okay, if the waters rise, we will face it and you will have to face this nonviolent action at a very different moral plane, because it was only to this that we knew it was beyond the government's capacity to respond. Otherwise they had guns and lathis, they had bombs—(whatever they wanted)—they even had the military at their disposal. . . ."[65] The jal samarpan tactic was not implemented that year since the waters did not reach the village of Manibeli, but the Andolan's commitment to it was not retracted.

Construction on the Sardar Sarovar Project continued but slipped behind schedule because the resettlement program was not being implemented and the environmental safeguards had not been completed. Relocation was not only proceeding slowly as a result of the domestic resistance to it, but also because some of the villagers who had earlier accepted resettlement packages had begun to return to their original villages due to the unacceptable conditions they had found at the resettlement sites.[66] Moreover, as of 1992, the comprehensive environmental plan that had been a condition of the environmental clearance by India's MOEF in 1987, and the environmental studies requested by the World Bank to be completed by November 1988, still had not been prepared and thus not submitted. [67] In fact, the MOEF's conditional environmental clearance had officially expired in 1989 due to the failure of project authorities to meet the requirements of the approval.

The Andolan and its supporters both inside and outside of India continued to focus on these failures while human rights nongovernmental organizations focusing on civil rights and liberties joined the struggle more assertively by preparing various reports on the repressive tactics being employed by dam proponents in the Narmada Valley. For example, sympathetic officials in the Ministry of Environment and Forests and nongovernmental members of the two sub-groups of the Narmada Control Authority highlighted the incomplete plans for environmental mitigation, as well as resettlement and rehabilitation. Moreover, Indian nongovernmental organizations, like the Peoples Union for Civil Rights and foreign nongovernmental organizations like Human Rights Watch/Asia published scathing reports documenting human rights violations by domestic authorities and argued that the World Bank had disregarded its own policies.[68]

THE WORLD BANK'S INDEPENDENT REVIEW

The central focus of the conflicts over the Narmada Projects at this point was the investigation of the World Bank's Independent Review of the Sardar Sarovar Major Dam project. The Independent Review team began its work in September 1991 and issued a 363-page report in June 1992, the same time that the Human Rights Watch/Asia Report was published. Throughout this period, the information relay among the various Indian and foreign allies in the transnational anti-Narmada coalition was fully mobilized to insure that all positive statements made by big dam proponents were countered immediately and effectively. After ten months of extensive investigation in which the review team met with Bank staff, governmental and bureaucratic officials, project authorities, project-affected people, and representatives of foreign and domestic nongovernmental organizations, the Independent Review offered a number of findings and recommendations.[69]

Demonstrating the critical importance of progressively more authoritative globalizing human rights and environmental norms, the Independent Review summed up their analysis with the following reflections:

> We have found it difficult to separate our assessment of resettlement and rehabilitation and environmental protection from a consideration of the Sardar Sarovar Projects as a whole. The issues of human and environmental impact bear on virtually every aspect of large-scale development projects. Ecological realities must be acknowledged, and unless a project can be carried out in accordance with existing norms of human rights—norms espoused and endorsed by the Bank and many borrower countries—the project ought not to proceed. . . . The Bank must ensure that in projects it decides to support the principles giving priority to resettlement and environmental protection are faithfully observed. This is the only basis for truly sustainable development.[70]

These findings and conclusions were remarkably similar to the arguments made in the report prepared by the Hindu Nature Club and Kalpavriksh in 1984.

On resettlement and rehabilitation, the Independent Review concluded that: (1) the Bank and India both failed to adequately assess the human impacts of the projects, including the nature and scale of displacement; (2) there was a clear failure to consult the potentially affected people, resulting in substantial opposition by them supported by domestic and foreign nongovernmental organizations; and (3) India's poor historical record as well as the differing policy formulations and policy implementing abilities of the three State governments, undermined the likelihood that resettlement would ever be successfully completed. In terms of the environment, the team reported that: (1) measures to anticipate and mitigate environmental impacts were not properly considered in project design due to the lack of basic data and consultation with affected people; (2) the environmental work-plan, required by both the Bank under the 1985 credit and loan agreements and the federal Government of India's own environmental clearance, was still not available as of the summer of 1992; and (3) the lack of basin-wide planning and use of a pari passu, incremental strategy had resulted in studies and plans for a number of environmental impacts, not to mention mitigation efforts, falling far short of requirements.

The Independent Review further proposed that the Bank allocate time, money, and personnel to ensure that its own policies are implemented in all its large scale projects, particularly those involving big dams. With respect to the Narmada Projects, and the Sardar Sarovar Major Dam in particular, they suggested that the Bank should "step back" in order to conduct the necessary studies and incorporate the potentially affected people more effectively in the development process. Noting that the opposition had ripened into hostility, the authors suggested that implementation without these changes would be impossible except as a result of "unacceptable means."[71]

But the conflicts over the Narmada Projects were far from over. The Andolan and its supporters inside and outside of India were elated with the Independent Review's findings, while the projects' proponents from the Chamber of Commerce and farmers' lobby in Gujarat to Indian bureaucratic authorities and government officials to the World Bank's India Operations Department were highly critical. The Government of Gujarat quickly formulated a response to the report and sent it to the World Bank. In the response, the GOG criticized the Independent Review for ignoring the vast amount of studies, methodological innovations, and improved policies that had been incorporated into the Narmada Projects across the areas of hydrology, resettlement, environment, and integrated basin planning. In a thinly veiled attack on the ethnocentrism and elitism of the review team, the GOG also wrote that the Independent Review,

Ignored socio-economic approaches developed by painstaking studies of the tribal population to be affected by one of India's leading social science research organizations and superimposed a view of colonial and British anthropologists which treat tribals as noble savages to be protected from the mainstream. The Review made extremely uncultured remarks on India's religions and went out of the way to be sarcastic of the leadership of India's freedom movement.[72]

Responses from other project proponents in India were even more critical. The Gujarat Chamber of Commerce and Industry, for example, wrote an open letter to then President Lewis Preston of the World Bank stating that,

We have carefully considered the report of the Morse Commission on Sardar Sarovar Project and are constrained to state that it is not only motivated and malicious as well as misleading and distorted but also mischievous in intent. As a matter of fact, it is with deep regret and disappointment that we are pained to state that report is not only partial and prejudiced but also hollow and vague as well.[73]

Supporters of the Narmada Projects further argued that the Morse/Berger Report was part of a plot to keep India backward and poor.[74]

In reaction to the Review's findings, the India Operations Department of the World Bank also sent its own mission to the Narmada River Valley in July 1992. The Andolan did not welcome this Bank mission and protested against it in numerous villages, reissuing its proclamation of "our rule in our villages." Nevertheless the mission continued with its work and three separate teams conducted investigations over a two-week period. On August 7, 1992, M. A. Chitale, then Secretary of the Ministry of Water Resources, sent a detailed critique of the Independent Review that had obviously been supported and assisted by discussions held between the Bank mission and Indian authorities.[75] The mission subsequently published a report entitled "Next Steps," outlining an action plan for dealing with the recommendations of the Independent Review. The action plan outlined a six-month conditional period during which construction would be tied to improvements in environmental mitigation and resettlement efforts.

The domestic movement and nongovernmental organizations worldwide saw this proposal as a complete denial of the Independent Review's conclusions. Shripad Dharmadikary of the Andolan severely criticized the "Next Steps" document at the World Bank/IMF annual meeting in September 1992. He warned that repression would be intensified and that even the temporary submergence that might be caused by continued construction was a violation of the oustees constitutional and human rights.[76] The Narmada International Action Committee published its own open letter to President Preston of the World Bank in

The Financial Times on September 21, 1992. The full-page letter, which was signed by 250 nongovernmental organizations from 37 countries, demanded that the Bank stop funding the projects because "It would be one of the worst human and environmental disasters the World Bank has ever financed."[77] Letters, faxes, and e-mails from all over the world, including from Morse and Berger, stating the Independent Review's conclusions were being ignored, poured into the World Bank.[78] The Bank staff in turn drew up a series of extremely stiff benchmarks that would have to be satisfied by the various Indian governments within a six-month deadline for funding to be continued.[79]

However, six Executive Directors called for the immediate suspension of the projects, including the Bank's three largest funders: the United States, Japan, and Germany. Not surprisingly, these were also countries in which critics had played a sustained role in the transnational anti-Narmada coalition. The Bank's credibility had no doubt been severely damaged in the eyes of both the representatives of several of its member governments and non-state actors around the world, especially after the Indian State governments announced that they would not fulfill the new benchmarks. Indeed, as had been predicted by anti-dam activists, human rights violations including beatings, arrests, detentions, rapes, even killings escalated during the six-month period.[80]

All participants and observers knew the Bank's benchmarks would not and could not be met as the March 31, 1993, deadline grew closer and closer. Finally, the day before the deadline, and four days after a meeting of the Narmada Control Authority on March 26, 1993, at which the decision was made, the Indian Executive Director to the World Bank formally announced that India would voluntarily forego the remaining $170 million of the $450 million credits and loans that had been allocated for the Sardar Sarovar Project in 1985. India would also not seek the approximately $500 million in additional support from the World Bank for the other Narmada Project that was under negotiation. As the then Secretary of India's Ministry of Finance stated, "We took the decision (to refuse World Bank funds) because we did not want this issue to be politicised in international fora. There is no question of it being a face-saving measure on our part to preempt stoppage of Bank funding."[81]

DOMESTIC MOBILIZATION AND DEMOCRACY: THE STALLING
OF PROJECT CONSTRUCTION

The transnational anti-Narmada coalition had clearly achieved a powerful, and certainly unexpected, victory in their campaign to prevent construction of the Narmada Projects. As a direct consequence of the World Bank's withdrawal, international support for the Sardar Sarovar Project from most sources including bilateral aid from a number of European countries was no longer forthcoming. But many Indian authorities vowed to complete that project and the broader scheme with their own and, if necessary, other types of external resources. As B. N. Nayalawla, then irrigation advisor to the Planning Commission, proposed, "The Eighth Plan envisages a total outlay of Rs 22,000 crore on major and medium irrigation projects. This will be more than enough to fund this project without assistance from the World Bank."[82]

Other options for raising the funds necessary to complete the project were pursued. The GOG, not taking any chances because the process of transferring the Sardar Sarovar Project to federal control might run into political difficulties, initiated plans to sell bonds to nonresident Indian (NRI) Gujaratis living abroad. In fact, over the next few years a succession of top-level State officials and project authorities made numerous trips, primarily to the United Kingdom and the United States, to publicize the Narmada bond-issue to the NRI Gujarati communities in those countries.[83] The GOG also commissioned studies on the possibilities of privatizing the power component of the projects, but found it financially, legally, and bureaucratically prohibitive.

Nonetheless, in clear violation of the orders of Narmada Control Authority, Gujarat's Sardar Sarovar Construction Advisory Committee continued work throughout the summer monsoon months. By December of 1993 the GOG had begun pushing for the closure of the sluices and raising the height of the Sardar Sarovar Major Dam still further, claiming that resettlement and rehabilitation had been completed for all those to be displaced to a reservoir level of nearly 210 feet. There remained, however, considerable disagreement within the Narmada Control Authority on closure of the sluice gates.[84]

The domestic opposition forces quickly realized that the withdrawal of foreign funding from the projects meant that they could no longer use the direct support that foreign nongovernmental organizations had been able to provide during the transnational campaigns. Lobbying efforts by these nongovernmental organizations claiming that World Bank and other foreign donors were still legally responsible because of the role they had already played in the execution of the Narmada Projects had very little sticking power. In response, the Andolan, concerned individuals, and domestic nongovernmental organizations throughout India began lobbying the federal government to halt implementation and constitute the domestic review of the projects which had been promised as a condition for ending the fast conducted during the "Long March" of 1990.

State authorities, however, unleashed a wave of repression in the Narmada Valley against the villagers and activists to demonstrate their resolve to move ahead with implementation. In response, the Narmada Bachao Andolan announced that it would once again launch a monsoon "Satyagraha" or "nonviolent protest" at the village of Manibeli on June 10, 1993. In particular, a "Samarpit Dal" or "Save and Drown Squad," consisting of activists and villagers representing thousands of families who had resolved not to move out of the valley, pledged to drown if any piece of land was submerged due to construction on the projects or the closing of the sluice gates.[85] Medha Patkar also began a fast in mid-June 1993 to protest the atrocities being committed by officials and police, as well as to demand for the comprehensive review.

Indian authorities were once again compelled to negotiate. After fifteen days of the fast and continued criticism across India of the Narmada Projects, the federal Government's Minister of Water Resources, V. C. Shukla, offered to hold talks with representatives of the Andolan. In the letter, he agreed to the Andolan's three conditions: (1) inclusion of independent experts at the talks; (2) constitution of a comprehensive review; and (3) stoppage of all work on the projects.[86] The discussions were held on June 29th and 30th and it was agreed that an independent domestic review team would be established. A series of letters passed between the Ministry of Water Resources (MOWR) and the Andolan during the month of July through which negotiations over the members and terms of reference of the review team were conducted.[87] The MOWR subsequently constituted a Five-Member Group to reappraise the projects by an office memorandum on August 3, 1993.[88]

But the Andolan was not initially satisfied. It criticized the terms of reference of the Five-Member Group, because neither clear reference to a review of the Sardar Sarovar Project, specification of a time limit for the investigations, nor indications that the report would be made public had been stipulated. The Andolan thus vowed to continue its opposition against the projects and did not attend the preliminary meeting of the Five-Member Group held on August 5th. That same day, the Five-Member Group issued an appeal to the Andolan, while the Ministry of Water Resources issued an amended office memorandum announcing that the Group should submit its report within three months, that the report would be made public, and that the Group would review all issues related to the projects.[89]

After ensuring that the Five Member Group was sufficiently balanced in terms of its membership and that its terms of reference were broadened, the Andolan suspended the jal samarpan program and agreed to participate in the review process. However, the authority of the Five-Member Group was soon undermined. Five days later, the Minister of Water Resources observed during a question and answer session of India's Parliament that the NWDT's ruling of 1979 prohibited any review before the year 2024. On September 30, 1993, the Secretary of Water Resources sent a letter to the Five-Member Group stating that the Narmada Projects were not under review and implementation would continue.[90] Moreover, not only did project authorities and State government officials refuse to participate in the review group's proceedings, but they also petitioned the Gujarat High Court, challenging the constitutionality of the review. The GOG further continued to formulate plans to raise the Sardar Sarovar Dam height to 240 feet before the next monsoon season.[91]

Indeed, construction on the Sardar Sarovar Project did continue. Between June and December 1993, while the Five-Member Group was conducting its investigations, the Sardar Sarovar Construction Advisory Committee had raised the height of the Sardar Sarovar dam to approximately 210 feet in direct violation of NCA orders. Gujarat project authorities then closed the sluice gates of the dam on February 24, 1994, even though the GOI's Secretary of the MOEF stated that this action was in violation of decisions made by the NCA. Lok Adhikar Sangh, a Gujarat-based human rights nongovernmental organization, immediately filed a public interest lawsuit on behalf of the soon-to-be-displaced tribal people in the Gujarat High Court. The High Court ruled in May that under certain conditions, construction could proceed and that the review group's report could not be made public until the court completed its deliberations on that matter.[92]

As a result, the Andolan and its supporters renewed their mobilization and lobbying activities at the dam site, throughout the Narmada Valley, at the capitals of the States and in New Delhi. The pledge to drown rather than be displaced was also reissued. As a tribal leader later expressed to me, "We decided that we would cling to the land like a baby clings to its mother. Our decision to be submerged was the logical consequence of our long struggle to save the Narmada and to force a rethinking of development in India and around the world."[93]

A series of activities refocused international attention on the continuing struggle against the projects. A proclamation calling for a moratorium on all World Bank funding for large dams, entitled the Manibeli Declaration "in honor of the heroic resistance of the people of India's Narmada River Valley," was drafted and signed by hundreds of nongovernmental organizations from all over the world. It was spotlighted in June 1994 as part of the ongoing "Fifty Years Is Enough" campaign against the World Bank (see chapter 6). At the same time, simultaneous protests against the projects were held at sites in the Narmada Valley, Bombay, Pune, Delhi, London, and Washington, D.C. Then, on July 21 the Five-Member Group submitted its report to GOI's Ministry of Water Resources. But the report was not released due to the Gujarat High Court's ruling to keep it secret until it passed judgment on the matter.

The Andolan and its domestic supporters responded by raising the stakes still higher when they filed two cases in the Supreme Court of India in May 1994.

The first asked for the overruling of the orders of the Gujarat High Court, to make the report of the Five-Member Group public and to place a stay order on construction of the dam. The second, and more comprehensive, public interest litigation claimed that the execution of the Sardar Sarovar Project was a violation of the fundamental rights to life and livelihood of the people to be displaced, under articles 14 and 21 of the Constitution of India:

> The grievance of the Petitioner in this Petition is that a large number of persons, mostly tribals and other marginalised sections of society, are being forcibly displaced and uprooted from homes and lands on account of this Project without giving them any opportunity to be heard and without properly compensating or resettling them and without even properly explaining to them the nature of the Project and seeking their participation. . . . Moreover, the Respondents are going ahead with the construction of the projects without even having completed the studies regarding the various effects . . . and are in the process violating a large number of stipulations of the Narmada Water Disputes Tribunal (NWDT). . . . The Ministry of Environment and Forests and the Planning Commission.[94]

Furthermore, the writ petition asked the Supreme Court to restrain project authorities from proceeding with construction until a newly appointed body comprehensively reexamined the projects, and to ensure that local people where represented on all decision-making bodies related to the Narmada Projects.

The Supreme Court took up both cases in August and September of 1994, while mobilization and lobbying activities continued all across India. On November 21, Medha Patkar and three others went on an indefinite fast in Bhopal, the capital of Madhya Pradesh, to pressure the GOMP to demand that the GOG stop work on the project. After twenty-six days, the GOMP agreed to the fasters' conditions. On December 13, while the fast was still on, the Supreme Court ruled to overturn the Gujarat High Court's order and make the report of the Five-Member Group public. The report itself was highly critical of project authorities and implementation, although it did not recommend stoppage of work. After hearing additional arguments from the petitioners and respondents, the Supreme Court ordered on January 24, 1995, that the Five-Member Group should prepare an additional report on the hydrology of the Narmada River, height of the Sardar Sarovar Dam, and resettlement and the environment impacts.[95] The Nar-mada Water Disputes Tribunal had ruled on all of these issues, except the environment, more than fifteen years earlier in its 1979 award.

The Andolan received some support from the Five Member Group's "Further Report," submitted three months later, on April 17, 1995. The majority of the members concluded that the projects could be implemented only under a number of conditions. These included (1) completion of all studies and plans on re-settlement, as well as environmental impacts before any further construction; (2) full participation of local people in the projects with mechanisms for easy access to information— holding periodic public hearings in the Narmada Valley, establishing an independent committee to hear grievances of the project affected peoples, and including representatives of the people in all decision making bodies; and (3) reorganization of the bureaucratic framework executing the projects to ensure greater efficiency and accountability. One member of the group—the very same Secretary of Water Resources who had pressured the MOEF to grant a conditional environmental clearance for the project in 1987—even filed a dissenting opinion, arguing that it did not seem likely given the historical evidence and ground conditions that authorities were capable of ever fulfilling the social and environmental requirements for the projects to be considered a success.[96]

The conflicts then returned to one of India's critical democratic institutions—the Supreme Court. The justices allowed the petitioners and respondents four weeks to submit additional materials with respect to the case after reading the report in July 1995. The federal Government of India, the four State Governments, and the Narmada Bachao Andolan and its supporters submitted over thirty volumes of documents and affidavits. After numerous delays, the full hearings finally began in November 1995 and subsequent hearings were held in January and April of 1996, during which time a stay on all construction was mandated. The central arguments of the Andolan, as noted previously, were that the projects were violating the fundamental rights of the people under India's Constitution and that authorities had not fulfilled the conditions of various domestic and international norms, procedures, and agreements with respect to re-settlement and the environment. The Andolan further claimed that the financial and economic costs of the projects were far greater than the benefits and thus it was in India's interest to halt construction on the projects.[97]

The State and federal governments, for their part, pleaded that they had attempted to fulfill these obligations in good faith. Due to unwarranted agitation by anti-dam actors and ground-level complexities, they had not executed all such requirements but would do so as quickly as possible. The GOG and GOI, in particular, claimed that the projects were the most comprehensively examined development effort in India's history and that the NWDT award of 1979 disallowed any reappraisal of the projects until the year 2024. The Supreme Court therefore had no legal basis upon which to re-adjudicate the matter.

The dam proponents' case was weakened when the conflict that had bedeviled the Narmada Projects ever since the 1950s—between the GOG which wanted to proceed with building the Sardar Sarovar Dam to a height of 455 feet and the GOMP which still argued that the height could be lowered to 436 feet in order to reduce displacement effects without diminishing irrigation and power benefits—resurfaced. But the GOMP would not have been so strong with its objection had it not been under continued pressure from grassroots mobilization and public criticism in the Madhya Pradesh areas of the Narmada Valley to be affected by the Sardar Sarovar Project.

The Supreme Court initially responded by stating that the question of the development value of the project was not under its jurisdiction but that of elected and bureaucratic authorities in India's federal democratic system. However, the justices did find that the fundamental rights of people under India's Constitution were clearly at stake. Furthermore, due to the fact that Narmada Water Disputes Tribunal could never have imagined how dramatically the institutional context would change in terms of resettlement at the time it issued its award, the Supreme Court could rule on these matters. Finally, the justices stated that the nonfulfillment of various environmental agreements and procedures by the project authorities and government officials was in clear violation of Indian and international law.

Consequently, given their authority in India's democratic system, the Supreme Court ordered an indefinite stay on implementation of the Sardar Sarovar Project. The justices also ruled that the parties would have to decide among themselves with regard to the ultimate scope of the projects via the institutional machinery established over the years since the NWDT award. No matter what came of those deliberations, in a strongly worded statement given at the February 1997 hearings on the case, the Supreme Court judged that the State and federal governments could not move ahead with any construction on the projects as long as resettlement and rehabilitation of project affected persons was not executed and as long as required studies, plans, and mitigation efforts on environmental impacts were not completed.

Gujarat authorities, however, continued to build the Sardar Sarovar Project without complying with the Supreme Court's February 1997 judgment. By this time, moreover, Hindu fundamentalist groups had grown in strength and the right wing Bharatiya Janata Party (BJP) had been elected to numerous governments in India including in the State of Gujarat. As suggested in chapter 2, the increasingly authoritarian nature of India's political system, along with mounting counter-mobilization by big dam proponents, and growing attempts to privatize big dam projects began to curtail the effectiveness of big dam opponents, including the Narmada Bachao Andolan and its transnational allies.[98]

Nevertheless, for more than 15 years through the late 1990s, the Save the Narmada Movement and its allies across the country and around the world had dramatically reformed and considerably slowed construction of the Sardar Sarovar

Project, as well as unexpectedly altered the implementation of the broader development scheme that involved dozens of more big dams to be built in the Narmada Valley. The campaign that focused predominantly on the Sardar Sarovar Major Dam component of the Narmada Projects along with the broader basin-wide and country-wide social movements contesting big dams also contributed to discernibly altering a range of development strategies, policies, procedures and structures in the country. And perhaps as importantly, these actors powerfully challenged the broader vision of development that had been hegemonic in India since Independence in 1947.

THE NARMADA CAMPAIGNS IN COMPARATIVE AND TRANSNATIONAL PERSPECTIVE

The two campaigns contesting the Narmada Projects during the 1980s and 1990s thus demonstrate the potential impact of transnationally allied advocacy and globalizing norms on development processes and outcomes. The first transnational reform coalition was effective in altering the resettlement aspects of the Sardar Sarovar Project by linking the lobbying efforts of domestic groups and transnational nongovernmental organizations to emergent globalizing norms on resettlement and the rights of indigenous/tribal peoples. The second transnational anti-dam coalition was effective in stalling construction on the Sardar Sarovar Project (and thereby affecting dynamics of other Narmada Projects) by linking an even more organized and sustained domestic social movement with transnationally allied advocacy efforts to a combination of transnational (international and domestic) norms on resettlement, tribal peoples, human rights (broadly construed), and especially the environment.[99]

The political opportunity structures available to transnationally allied actors were discernibly altered by the gradual institutionalization of globalizing norms both at the international and domestic levels. In the case of India's Narmada Projects, the creation of environmental procedures and structures at the World Bank and India, for example, clearly opened up new avenues of access and contestation for the members of the transnational coalitions that would not have existed otherwise. In other cases, other types of globalizing norms on issues ranging from human rights to indigenous people, gender justice to anti-corruption have been or could become important.

The impact of the dialectical interactions between transnationally allied advocacy and globalizing norms in both Narmada campaigns clearly depended on the existence of grassroots mobilization and sustained domestic opposition in the context of a relatively democratic regime in India. Domestic social mobilization dramatically increased the costs (both political and economic) faced by Indian pro-dam actors and their transnational allies in forging ahead with the projects. It also added to the legitimacy of the lobbying activities of members of the transnationally allied critics who were not directly affected by the projects.

Furthermore India's political system, at least through the mid-1990s, offered a set of democratic political opportunity structures that gave domestic actors the right to organize and mobilize, the ability to forge coalitions with like-minded foreign actors, and access to large quantities of information via a considerably free press, relative access to decision-making authorities, and a legal system in which the courts could be petitioned to hold state and non-state actors accountable to Indian and international norms and institutions.

Indeed, as the next chapter will reinforce, while transnational linkages and the global spread of supportive norms and institutions have become progressively ubiquitous around the world since the 1970s, it is the two domestic level factors of democratic political opportunity structures and organized domestic opposition that can be crucial to dynamics around big dams and development.

Finally, as will become evident in the concluding chapter of this book, the transnational campaigns around the Narmada Projects were as vital to changes outside of India and at the international level as they were to domestic dynamics in the country. The second transnational anti-Narmada campaign played an especially crucial role in the formation of a transnational anti-dam network, and the establishment of an unprecedented World Commission on Dams. The transnational anti-Narmada campaign was also critical to the linked anti-multilateral development agencies' campaign that contributed to major reforms at the World Bank along with other development donor organizations around the world. Thus the transnational struggles over India's Narmada Project both shaped and were shaped by the changing dynamics of big dam building and development from the local to international levels.

❖

CHAPTER FIVE

Dams, Democracy, and Development in Comparative Perspective

Conflicts over big dams have proliferated and intensified throughout the third world since the 1970s. The mounting opposition to these projects correlates with increasing grassroots mobilization and growing numbers of nongovernmental advocacy organizations in domestic civil societies in such countries as India, Brazil, South Africa, Indonesia, and even China, among others. Independent domestic resistance to big dam building by itself has generally been unsuccessful at substantially reforming or halting the construction of these projects in these societies. But when these domestic groups are linked transnationally to allied critics from around the world, directly and often times even indirectly, they have been greatly empowered and more likely to alter both decision-making processes and development dynamics.

Big dam opponents have also been empowered by the global spread and deepening of particular types of norms at the international and domestic levels. The more democratic decision-making processes are, the more political opportunities these actors have to organize and mobilize, with less fear of being physically repressed. Democratic institutions also allow these actors multiple mechanisms in order to gain access to and leverage decision-makers, including a free press, competing political parties, legislative assemblies, and the courts. Thus, the international and domestic institutionalization of more democratic political regimes, along with the global spread of human rights, indigenous peoples, and environmental norms in the form of bureaucratic agencies and legal procedures, has provided additional openings and resources that nongovernmental organizations and social movements could use in order to alter development policies and projects, such as big dams.[1]

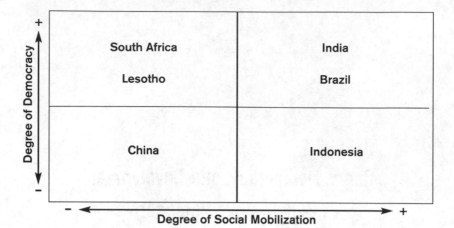

Figure 5.1. Country Case Selection: Democracy and Mobilization

In sum, the coordinated lobbying of transnationally allied nongovernmental organizations and social movements, the global institutionalization of legitimating norms and principles plus democratization of decision-making and domestic social mobilization have profoundly changed the dynamics of big dam building—and thus development—since the 1970s. This chapter offers a comparative examination of big dam building in four third world contexts in order to empirically assess this general theoretical argument (which was derived and refined from the cross-sectional and cross-temporal analysis of these dynamics in India generally and the Narmada Projects specifically conducted in chapters 2 through 4). The comparative cases were chosen to demonstrate the range of outcomes based on variation in two domestic level factors: degree of democracy and degree of social mobilization (see figure 5.1).

The central hypotheses examined in this chapter are that the presence of democratic institutions and the level of social mobilization domestically mediate the effects on big dam building in the third world of transnationally allied opposition and institutionalization of norms about the environment, human rights, and indigenous peoples. Thus, as depicted in figure 5.2, major reforms and significant declines in big dam building conditioned by transnationally allied actors and globally spreading norms are most likely to be found in those third world contexts with greater degrees of democracy and higher levels of civil society mobilization/organization. Change would be most unlikely in third world contexts that have stronger authoritarian regimes and weaker civil societies. The cases that have either democracy or social mobilization but not both would probably demonstrate an intermediate level of reform or decline in big dam building.

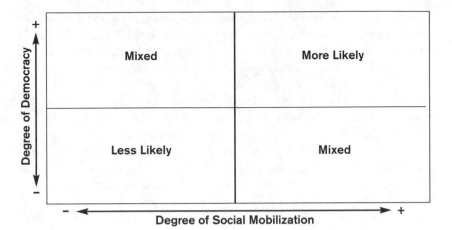

Figure 5.2. Impact of Domestic Democracy and Social Mobilization

Brazil, Indonesia, China, and South Africa had successful records of big dam building similar to that of India from the 1950s through the 1970s.[2] Since the late 1970s, the increasing organization of grassroots social mobilization linked to transnationally allied opposition and the domestic adoption of norms about the environment, human rights, and indigenous peoples has resulted in the declining ability to construct big dams in Brazil. But the degree of change was somewhat less than in India during the last quarter of the 20th century because Brazil's transition to democracy only began in the 1980s. However, with the continuing deepening of democracy in Brazil through the late 1990s and the increasingly authoritarian nature of India's political system, the outcomes may be more dramatic over the long term in the former rather than the latter country.

By contrast, big dam building dynamics have not been altered as dramatically in Indonesia and South Africa. Although transnational coalitions of nongovernmental organizations and the presence of grassroots anti-dam mobilization resulted in some reforms in Indonesia, the persistence of an authoritarian regime backed by the military through the 1990s gave government officials, project authorities, and other big dam proponents a relatively unchecked ability to repress big dam opponents. The subversion of not only established legal procedures, but also newer norms and institutions on the environment, further inhibited the impact of struggles around big dams there. Comparison with yet another case, that of South Africa, demonstrates that the impact of criticism and opposition at the international level, the existence of supportive global norms, and even domestic democratization are going to be limited in the absence of strong domestic social mobilization.

In the case of China, both the absence of organized, sustained domestic resistance and the presence of a powerful authoritarian regime have resulted in few possibilities for transnational links to be formed and for global norms to be consequential. Thus, far less change in the dynamics of big dam building has occurred in China than in any of the other cases to be examined. Big dam building dynamics in China thus provides supporting evidence as a negative case that can be accounted for by the broader theoretical argument of this book.

In this chapter, I first analyze the transformation in big dam building dynamics that has occurred in Brazil and elaborate on this case more thoroughly than the others because it is where the shift in norms and practices has occurred most discernibly. The other cases—China in particular—exemplify far less or very little change and thus require less detailed accounts. After providing evidence of the general trends and patterns in big dam building in each country, I focus primarily on specific critical and often highly symbolic big dam projects and conflicts that are both indicative and generative of broader development dynamics. While the following analysis is not intended to empirically confirm the general theoretical argument of this study, it does add further support to its potential descriptive validity and explanatory power.

BRAZIL: DOMESTIC MOBILIZATION WITH DEMOCRATIZATION

Big dam building, primarily for generating hydroelectric power and flood control, constituted a central component of Brazil's development strategy during much of the 20th century. Brazil built by far the largest number of big dams of any country in Latin America and ranked among the top ten in the world in this category. But, according to most reliable statistics available to date, there has been a dramatic decline in big dam building in Brazil over the last three decades.[3] While 100 big dams were built during the 1950s and 103 during the 1960s, this rate dramatically dropped to 91 in the 1970s, 60 in the 1980s, and to less than 30 in the 1990s.

Big dam building efforts were directly linked to Brazil's top-down, technocratic, state-directed development model, which was first adopted partly in reaction to the international economic crisis of 1929. This statist development strategy prioritized the promotion of economic growth through the intensive exploitation of Brazil's vast natural resources. Thus, by 1937, Getulio Vargas, the first Brazilian president to ever visit the Amazon, announced, "I have come with the purpose of seeing to the practical possibilities of putting into execution a plan for the systematic exploitation of the wealth and the economic development" of the region.[4] As Boliver Lamounier notes, "subsequent governments assumed an increasingly direct role in the economic sphere. Starting with the Volta Redonda steel complex, in 1942, state and mixed enterprises were created to foster industrial infrastructure."[5] The huge Furnas Dam on Brazil's Rio Grande—the largest such project in Latin America at the time—was initiated in 1958, as part of Brazil's industrial push, with a loan from the World Bank. It was completed in conjunction with the Canadian-owned São Paulo Light Company that not only bought stock in the state enterprise that would operate the dam, but also agreed to buy every megawatt of power produced.[6]

Brazil's statist and capital intensive development model was further consolidated after the coup of 1964, which displaced the populist João Goulart administration and ushered in an authoritarian regime based on an alliance between military hard-liners and economic technocrats.[7] Brazil's bureaucratic authoritarian regime, as Mark C. MacDonald suggests, "promoted the ideology of 'Brasil Grande,' initiating enormous projects such as the Trans-Amazon Highway, the Itaipu Dam, and the Carajas mining project," which "assumed high and steadily increasing rates of energy consumption over several decades. Massive expenditures in energy production would become one of the principal areas for state investment as hydroelectric projects took on a new significance."[8]

A comprehensive restructuring and expansion of the energy sector was initiated, beginning in the early years of the newly ascendant military regime. Eletrobrás, the national electricity holding company under the auspices of the federal Ministry of Mines and Energy, quickly conducted a study to determine Brazil's exploitable energy resources. Eletrobrás created Eletrosul in 1965 to implement its energy development program in the South of Brazil and Eletronorte to do the same in the Northern regions in 1973. By the 1980s, 18 concessionary firms including a majority of whose principal shareholders were state governments were coordinated by Eletrobrás. This tremendous growth in size, organizational complexity, and stature was fueled by the support given to the energy sector by the regime leadership, as well as by the increasing involvement of foreign donors (led by the World Bank) in Brazil's development.

Although the World Bank had mostly refused to give loans to the previous populist and elected governments of Brazil, after the 1964 coup, lending rose to $73 million per year by the late 1960s and to almost $500 million annually by the mid-1970s.[9] Moreover, as Stephen Schwartzman and Michelle Malone note, "In the early 1960s, Brazil, faced with the increasing costs of foreign oil and the vast expenses of hydroelectric plants needed to generate energy, turned to the MDBs for energy sector financing." Between 1965 and 1985, the World Bank invested more than $3 billion in the Brazilian energy sector, which represented more than 25 percent of the total amount the Bank lent to the country and approximately 10 percent of the Bank's total financial involvement in Latin America and the

Caribbean during that period. The Inter-American Development Bank also provided Brazil with almost $2 billion for energy development during those two decades.[10]

The dramatic results of the authoritarian regime's development strategy, and the priority placed on hydroelectric energy expansion, is demonstrated by the massive big dam building effort that was executed in all regions of Brazil. The first two hydroelectric dam projects in the Amazon were quickly built under the auspices of the new military federal and state governments in the late 1960s. In order to more rapidly exploit the tremendous energy potential of the Amazon, Eletronorte planned numerous hydropower big dams during the 1970s and projected that all the economically justified sites in the region would be developed by 1990.[11] But, as we shall see, this ambitious target would not be achieved even by the late 1990s and for the foreseeable future into the 21st century.

Electronorte's plans included a massive dam building effort targeted for the Xingu River Basin for which a study conducted between 1975 and 1979 identified 47 potential sites.[12] Construction of the gigantic 8,000 megawatt Tucuri Dam was commenced in 1976 and completed in 1984. The project resulted in the dislocation of 5,000 families, including members of two indigenous tribes located in the area. The 250 megawatt Balbina Dam on the Uatuma River was approved by the military regime in 1982 in an election year effort to shore up support by providing employment and commerce to the state of Pará. The dam flooded two villages in which many of the remaining Waimiri-Atroari indigenous peoples lived before its completion in 1988, and inundated 31 times more forest area per megawatt generated than Tucuruí.[13]

In fact very few, if any, of the big dams planned and constructed up until the 1980s involved the participation of peoples to be negatively affected by these projects, nor was adequate compensation given to those who were harmed or displaced. Construction on the World Bank funded Sobradinho Dam project in the northeast region's São Francisco Valley was initiated by CHESF (Companhia Energetica de São Francisco), another subsidiary established by Eletrobrás, in 1972. The 1050 megawatt dam project eventually dislocated an estimated 72,000 people and inundated more than 4,000 square kilometers of land; however, by all accounts displaced persons were not consulted prior to its sanctioning nor satisfactorily remunerated.

In 1976, one year before the sluice gates of the Sobradinho Dam were closed, work on the nearby 2,500-megawatt Itaparica major dam project commenced without notifying the peoples in its command area. By the time the sluice gates of Itaparica were shut in 1988, more than 44,000 people had been displaced and a total of nearly 118,000 people had been negatively affected. Planning for the 5,000-megawatt Xingo project and the 600-megawatt Pedra do Cavalo project was also initiated in similar top-down and insulated fashion by Eletronorte while Sobradinho and Itaparica were being constructed.[14]

Perhaps the most symbolic, and certainly the grandest, of the big dam projects built during the authoritarian regime's and Eletrobrás's massive energy development drive was Itaipu. Brazil and Paraguay negotiated an agreement in the early 1970s to construct the Itaipu major dam on the Parana River in southern Brazil, near the border between the two countries.[15] When it was finally completed in 1982 with a total cost of more than $20 billion, it became the world's biggest hydropower project with a generating capacity of 12,600 megawatts! EBI (Entidad Binacional de Itaipu), the joint Brazilian-Paraguayan public enterprise established to build and operate the project, borrowed the vast amount of the money it needed from private foreign banks. The project ultimately displaced more than 42,000 people and inundated over 1 million square kilometers of land.[16]

But Itaipu was not the only big dam project initiated by Eletrobrás in its southern expansion under Brazil's bureaucratic authoritarian regime. During the 1970s, three other major 1,000+-megawatt dam projects were also completed on the adjoining Iguacu River near Itaipu.[17] By the end of the decade, moreover, Eletrosul proposed constructing twenty-five dams on the Uruguai River, primarily in the southeastern states of Santa Catarina and Rio Grande do Sul. The original estimate of the population to be displaced by Eletrosul put the figure at nearly 40,000 people and the projected extent of inundation covered virtually all of the 2,000-kilometer-long Uruguai River. Prior to the publication of a revised study outlining the twenty-five dam scheme in 1979, neither the region's residents nor most of the state-level officials were aware that an initiative of this scale was envisioned.

Domestic criticism emerged and gradually mounted against the massive big dam building initiative under way during the 1970s. As Patrick McCully writes, "People harmed by major projects such as Sobradinho, Itaipu, Itaparica and Tucuruí fought for better compensation through marches, dam-site occupations and other forms of civil disobedience, but failed to win significant concessions."[18] The opposition, which focused initially on improving the compensation and resettlement packages of those to be negatively affected and not on halting dam building altogether was, however, generally quashed by pro-dam actors under Brazil's dictatorship, which limited freedom to organize and access to information.

The military authoritarian regime inhibited and often outright repressed opposition to its top-down, technocratic, economic growth-oriented and natural resource intensive development model, particularly during the initial decade of its reign. Thus, when criticism of big dam building did exist, it could not be expressed fully, let alone be organized and mobilized to its fullest extent. Bolivar Lamounier describes the political context—and thus the political opportunity structure facing civil society—at the time in the following manner: "The legislature, political parties, the judiciary: all of these came out clearly weakened vis-à-vis the executive (which in fact meant the military). With the military in power after 1964, this trend became far more serious. First came the arrests, proscriptions, and similar measures designed to curb opposition and promote societal demobilization. From 1968 to 1974, confronting armed underground movements, the regime adopted widespread censorship and all sorts of cover-up repressive practices."[19]

As a result, affected peoples, critics, and opponents of big dams often joined the struggle for democratization that was being waged by various other groups and movements in civil society.[20] The protracted and imposed political liberalization and re-democratization process orchestrated by the military regime from 1974 to 1985 gradually opened up greater opportunities for the activity of civil society organizations in Brazil. For example, as Stephen P. Mumme and Edward Korzetz state,

> An incipient environmental movement emerged in the early 1970s in response to the modest liberalization by the military government of Ernesto Geisal . . . which was accompanied by the formation of a number of ENGOs (environmental non-governmental organizations) between 1974 and 1981, including several by the late 1980s.[21]

It was during the latter period that the growth and increasing organization of grassroots mobilization by peoples affected by the dam similarly contributed to the declining ability to build big dams or in substantial reforms in the formulation of these projects in Brazil.

Indeed a large number of social and political movements emerged and progressively became more organized in Brazil during the late 1970s and the beginning of the 1980s. These movements include the "Movimento dos Sem Terra" (MST), organized in 1984, the "Partido dos Trabalhadores" (PT) in 1980, and the "Central Unica dos Trabalhadores" (CUT) in 1983. In addition, during this time, a great number of urban social movements, like the movements in "Favelas," were also born. These movements and organizations constituted the broader "rebirth of civil society" in the country and were crucial to the Brazilian re-democratization process.[22]

Re-democratization of the Brazilian regime not only contributed to the increasing ability of big dam opponents to organize and effectively challenge state institutions and practices. It also opened up greater space for the progressive domestic institutionalization of globalizing norms about the environment, human rights, and indigenous peoples that had already begun to seep into Brazilian politics and that were essential to altering the dynamics of big dam building in the country from the 1970s on. For example, the Indian Statute Law #6.001, based on the norms promoted in the Geneva Convention on the Protection of Indigenous Populations, was enacted in 1973. As a result of the new regulation, the National Indian Foundation (FUNAI), which had been established in 1967 and located in the federal Ministry of the Interior, was required and increasingly pressured to reorient its work toward protecting indigenous peoples who had been victimized by the implementation of Brazil's top-down, technocratic, and natural resource exploiting development model.

Norms for the protection of indigenous peoples became further consolidated in Brazil during the 1970s and 1980s, in large part due to increasing transnationally allied pressure from both external actors—especially the emergent transnational nongovernmental organizations promoting the protection of indigenous peoples identified in chapter 1—and the growing domestic Indian movements and pro-Indian nongovernmental organizations within the country.[23] The culmination of this transformation occurred with the inclusion of several clauses on the rights of tribal peoples in the democratic Brazilian Constitution passed in 1988. Article 231 of the Constitution is emblematic of this change. It required that the social organization, customs, languages, creeds, traditions, and lands of Indians be protected, and also mandated authorization from the National Congress and consultation with the affected Indigenous groups when water resources were to be developed in tribal areas. Subsequently, an Indian Lands Environmental Service (SEMATI) was established by FUNAI in 1989.[24]

The recognition of an intimate linkage between indigenous peoples and natural resources also increased along with the progressive domestic promotion and institutionalization of environmental norms in Brazil from the 1970s on. In fact, attitudes toward the environment in Brazil dramatically changed within a decade after the 1972 United Nations Conference on the Human Environment, from "the view that environmental activities are luxuries impeding development . . ." toward "the view that since prevention costs less than cure, developing countries would be wise to allocate scarce resources with more environmental prudence."[25] Similar to the case of India, the government quickly created a Special Environmental Secretariat (SEMA) in October of 1973 within the Ministry of Interior following the Stockholm Conference. Although SEMA did not wield a great amount of authority early on, it did successfully formulate a formal and comprehensive National Environmental Protection Act with assistance from Brazil's emerging environmental nongovernmental organizations that was enacted into law in 1981.

Environmental norms were further institutionalized domestically in Brazil with the establishment of the National Council on the Environment (CONAMA) in 1985 that was a partial consequence of the 1981 national environmental policy. As with other such government bodies established at the time in Brazil, CONAMA consisted of members from federal and state agencies as well as civil society—including individuals from Brazil's burgeoning environmental movement. As Eduardo Viola notes, "Once under way, CONAMA began producing crucial resolutions for Brazilian environmental protection, such as requirement for environmental impact assessments on new projects (industries, roads, airports, dams, and so on). Such assessments would include public hearings as a mechanism for social participation in decision making." Environmental agencies and regulations were also instituted at the state-level in the south and southeastern regions of the country during this period.[26]

After the passage of the new democratic constitution in 1988, the Brazilian Association of Environmental State Agencies was founded to further strengthen federal-state coordination of environmental policy formulation and implementation. The following year, the government merged SEMA and other federal agencies dealing with the environment under a single organization—the Institute for the Environment and Natural Resources (IBAMA). IBAMA was placed under a new Secretariat of the Environment, created by the first President (Fernando Collor de Mello) to be elected in direct and free elections in Brazil after 1960. The first head of the Environment Secretariat was Jose Lutzenberger, one of the founders of Brazilian environmentalism during the 1970s; he was given the task of preparing Brazil to host the United Nations Conference on the Environment and Development in 1992.

Thus, in less than two decades, a virtual transformation in Brazil's formal structures and procedures had occurred. This included a transition from a military authoritarian regime to a constitutional democracy and (re-)prioritized human rights, as well as the establishment and progressive consolidation of bureaucratic agencies and rules for the protection of the environment and indigenous peoples within Brazil. Civil society organizing and social mobilization against big dam projects both contributed to, and was empowered by, these multiple and related processes of institutional change. As we shall see, transnationally allied nongovernmental organizations and social movements, often through their lobbying efforts vis-à-vis international organizations like the World Bank, were critical not only to these formal institutional innovations, but to the overall shift in big dam building practices in Brazil as well.

Domestic criticism of the negative effects of big dam projects in Brazil originated at the local and regional levels and gradually became more national and transnational over time. As with the protests against most other big dams until the 1980s, the campaign around the Itaipu Dam was focused on improving resettlement and compensation packages for negatively affected people. The Brazilian Lutheran and Catholic churches near Itaipu, along with the Pastoral Land Commission (CPT), began working with the peoples affected by the project in 1978. With this support, the Itaipu affected peoples quickly organized themselves into the Movement for Justice and Land (Movimento por Justica e Terra) and formed coalitions with various other rural workers associations in the southern state of Paraná. The grassroots movement orchestrated a series of protests and demonstrations, including a 57-day mobilization at the headquarters of the Itaipu Binational Company.

But the absence of transnationally allied support, the relative newness of indigenous peoples and environmental institutions in Brazil, as well as the campaign's lack of attention to the project's negative environmental costs and the devastation of the indigenous Guarani peoples, resulted in minimal reforms being obtained.[27] As in the cases of the Sobradinho Dam in the northeast and the Tucuruí Dam in the Amazon, the experience did nevertheless have a broader effect on spurring other anti-dam organizing in Brazil. For example, mobilization against the building of big dams on the adjacent Iguacu River emerged in 1983–1984 and was able to postpone the construction of the proposed Capanema hydroelectric project for almost a decade. That opposition soon evolved into the Regional Commission of Peoples Affected by Dams on the Iguacu River (CRABI), linked itself with other civil society organizations such as the rural workers unions in the area, and became a critical contributor to the formation of a country-wide anti-dam movement.[28]

The experience with Itaipu also affected the groups that coalesced to oppose the massive twenty-five dam scheme proposed by Eletrosul in 1977 for the Uruguai River Basin, in which the Machadinho and Itá Dams were central projects.[29] As with the other cases of peoples mobilization vis-à-vis such hydroelectric projects, various church organizations, rural unions, the CPT, and scholars from the nearby university were critical in the initial organizing efforts. With support from these actors, the organization, which later became the Regional Commission of People Affected by Dams (CRAB), was created in 1979 and evolved into one of the most organized movements in all of Brazil by the 1990s.

As with the other early big dam struggles in Brazil, the communities to be affected initially supported the building of the Uruguai River Basin Projects and were primarily concerned with receiving adequate and timely indemnification of their losses. But perceptions soon changed from a combination of factors including the lack of information on the proposed dam projects made available to local people by authorities and the stories that spread of the plight of peoples affected by other big dams. By 1981, consistent interaction with the peoples affected by the Itaipu and other completed projects in Southern Brazil was established and, in 1983 and 1984, CRAB leaders screened a video on the Itaipu experience throughout the Uruguai River Valley. Eletrosul's failure to satisfactorily answer the questions of CRAB during a meeting held in May 1982, and again at a statewide conference organized by the state legislative assembly of Rio Grande do Sul in September 1983, further contributed to the increasing suspicions of the region's inhabitants.

Given the increasing political openness available from the ongoing re-democratization process in Brazil, CRAB soon evolved into a powerful regional movement. Initially, a highly successful petition drive gathered over one million signatures against the projects. CRAB then established local and municipal offices in the areas of the Machadinho and Itá Dams with the assistance of the local rural unions and other nongovernmental actors, such as the Federation of Agricultural Workers (FETAG). Large statewide public conferences (Encontros Interestaduais) attended by thousands of people were organized. In 1984, an Executive Committee was formed to coordinate the movement's activities across the region. A Secretariat with a full-time director was set up to execute administrative functions, including the publication of the movement's periodical, "A Enchente do Uruguai," which not only publicized the negative social effects but increasingly also the environmental costs of the dam projects. CRAB even instituted a General Assembly in 1986 to ensure the accountability of the leadership to, and consistent participation in decision making of, the movement's members.

Despite the mounting and increasingly organized opposition, Eletrosul, Eletrobrás, and the federal Ministry of Mines and Energy refused to negotiate directly with CRAB or allow the peoples to be affected to participate in the planning efforts. Eletrosul's compensation package was unacceptable to CRAB, however, because it offered only cash for lost lands and assets on an individual basis, and indemnified only legal landholders that left the "sem-terra" or landless families out. Eletrosul attempted to bypass CRAB by co-opting local officials and leaders and even established a commission entitled "A Equipe de Justica e Trabalho" (The Team for Justice and Work) to undercut the movement's legitimacy in the region. But the commission lacked local support and eventually was dissolved. Nevertheless, CRAB's attempts to meet with high level officials continued to be consistently thwarted.

Given the persistently poor and even deteriorating relations with Eletrosul, and the subsequent lessons learned from experiences with big dams in the northeast, CRAB changed its reformist stance to a more strongly anti-dam position by the mid-1980s. Leaders of the movement traveled to the northeast and met with the affected peoples of the big dams being built in the Sao Francisco River Valley. At the Itaparica project area, they found that thousands of displaced peoples had not been compensated and that a comprehensive resettlement plan had not been drafted, despite the fact that the dam was nearly complete. The representatives of CRAB were advised by their northern counterparts to take proactive action by destroying prospective dam reservoir markers, presenting their own resettlement proposals, and compelling Eletrosul to sign an official accord before construction on the projects had progressed too far.

The anti-dam mobilization grew dramatically both in intensity and sophistication between 1984 and 1987. The affected people destroyed reservoir markers in one of a series of increasingly defiant acts. Confrontations between Eletrosul workers and the local residents became commonplace when the latter began surveying properties without the permission of landholders. Despite the fact that Brazil's federal Minister of Mines and Energy announced that work on the Machadinho and Itá Dams had been suspended until two appointed commissions formulated mitigation measures for the negative social effects of each, the movement continued its protests. On October 12, 1985, the first National Day of Protest Against the Construction of Dams and for Agrarian Reform was held. CRAB then withdrew its participation from the commissions after becoming disillusioned with their likely effectiveness. When Eletrosul's President failed to come to prearranged meetings with CRAB's leaders, the movement finally declared that no further work on the projects would be allowed.

The movement forged links with transnationally allied nongovernmental organizations and also participated in efforts to alter the ability of Brazilian officials to acquire electricity sector loans from international development agencies, such as the World Bank. In one case, the transnational lobbying campaign resulted in a series of conditions tied to World Bank loans for the electricity sector. The conditions required Brazilian officials to assess and mitigate the negative social and environmental effects of hydroelectric dams, such as Machadinho and Itá.[30] In another, when the Inter-American Development Bank (IADB) was considering a loan to specifically assist Eletrosul with the construction of hydroelectric dams on the Uruguai River, meetings between CRAB and IADB were arranged by the Environmental Defense Fund to discuss the various problems with the proposed projects. IADB decided not to fund Eletrosul at the time.

CRAB eventually succeeded in wresting concessions from project authorities in 1987, eight years after the movement had been initiated, but not without forming transnational coalitions or sustaining its vigorous grassroots mobilization. The latter efforts included demonstrations in which movement members and allies blocked highways for several hours. A proclamation was drafted that demanded that a full resettlement plan be drafted within a two-month period. When federal authorities sanctioned the initiation of work on the Itá dam, dozens of farmers in the area forcibly detained one of Eletrosul's chief engineers and compelled him to order a one-week stay on project work. The date for the resettlement plan was not met by Eletrosul, and the movement responded by capturing and detaining company functionaries. Finally, when 5,000 to 10,000 people surrounded the company's headquarters in the valley and announced that they were prepared to burn the building down, Eletrosul's president agreed to meet with CRAB.

But meetings held separately for the Machadinho and Itá Dams between CRAB and Eletrosul over the following years continued to prove conflictual. When project officials announced that negotiations had to be suspended due to financial constraints, CRAB initiated a campaign to permanently halt work on the Machadinho Dam. They were successful as work on that project was suspended in 1988. Those to be affected by the Itá Dam agreed that construction could continue as long as their demands with regards to resettlement and compensation for all, including the landless, were met by Eletrosul. But Eletrosul more often than not failed to do so. During the 1990s, a stalemate emerged which, for all practical purposes, meant that the anti-dam actors succeeded in their efforts to postpone construction until their demands were met.[31]

The implications of the conflicts over the big dam projects in the Uruguai River Basin were manifold. As Mark McDonald concludes:

> The changes in policy that occurred after 1986—direct negotiations with the antigidos, the establishment of accords concerning the displacement process, the resettlement of sem-terras—were the result of pressure on the company rather than social learning. Through protests, and demonstrations, through the media and public forums, through international alliances, and by capturing company workers and occupying offices, the movement successfully forced the debate over dams into the public arena and, in the end, had an impact on the policies of Eletrosul. . . . The democratic opening in Brazil over the last decade has allowed popular groups progressively more political space in which to organize and protest.[32]

In addition, as a powerful regional movement, CRAB contributed to the formation of a country-wide movement of peoples affected by dams by the end of 1980s that visibly altered the dynamics of big dam building in Brazil.

The negative consequences of big dams came to the public attention in the northeast even before it did in the south of Brazil, initially as a result of the shoddy and incomplete resettlement of the people displaced by the World Bank funded Sobradinho Dam. As early as 1976, the rural labor unions in the project's impact area initiated a campaign to improve the compensation packages that were being offered to the oustees. The sluice gates were closed just a year later, and thus the mobilization had very little impact. Many of the dam affected peoples filed court cases against CHESF, the regional electric power company that executed the project, with the assistance of the nongovernmental Pastoral Land Commission (CPT) in the area. The weakness of Brazilian courts under the persistent authoritarian regime, however, yielded few victories for the dam affected.[33]

The experience of the Sobradinho Project affected peoples was not lost on residents at nearby Itaparica Dam in the Sao Francisco River Valley, who began organizing almost immediately after construction started in 1976. By 1979, at the first meeting of residents from all the municipalities to be flooded by the dam, a decision was taken to demand lands for the lands to be inundated and a share of the benefits generated by the project. The Polo Sindical dos Trabalhadores Rurais do Submedio Sao Francisco was then established. It linked the rural workers unions of Petrolandia, Floresta, Itacuruba, Pernambuco, and eight other communities to be affected into a broad movement to reform the project. Over the subsequent eight years, numerous conferences and demonstrations were held and various other direct actions were taken. In 1982, for example, members of the movement attempted to halt work at the dam site but were forcibly removed by the police at the behest of CHESF. Three years later, the movement's leaders met with the federal Ministry of Mines and Energy, the Ministry of Interior, and with state officials of Bahia and Perambuco to present their case against the project.[34]

But it was not until pressure from transnationally allied nongovernmental organizations combined with the continued mobilization of grassroots groups that project authorities began to offer significant concessions. Nongovernmental organizations from outside Brazil, such as Environmental Defense Fund and Oxfam (the latter which had already been working on the ground with the Itaparica-affected), had increasingly become involved in a campaign to reform the destructive activities of the World Bank. When the Brazilian Government petitioned the Bank for a $500 million electricity sector loan in 1986, these and other nongovernmental organizations pressured the Bank to include explicit environmental and social conditions in the loan package. In particular, nongovernmental organizations lobbied the Bank to insist that the Brazilian Ministry of Mines and Energy sign an agreement with the Polo Sindical de Itaparica. As a result, an accord was signed in 1986 that guaranteed irrigated lands, houses with electricity, technical assistance, and the participation of the local peoples in all further resettlement efforts.[35]

With the domestic adoption of new procedures and creation of new agencies on the environment and indigenous peoples based on globalizing norms beginning in the 1970s, gradually more began to be implemented in these areas with regard to big dam building in Brazil and particularly in the Amazon. For example, Eletrobrás created an environmental department in 1975 at the urging of SEMA, which itself had been created only after the 1972 United Nations Conference on the Human Environment. Correspondingly, Eletronorte commissioned an environmental impact assessment of the Tucuruí Hydroelectric Dam Project in the Amazon two years later. But even in this case, when such a study was conducted, and most often they were not, the primary concern of the company was to salvage the timber to be lost by the reservoir inundation.[36] Neither the displacement of 5,000 non-indigenous families nor the flooding of the Pucuri Indian Reservation belonging to Parakana groups was considered an environmental concern at the time by Eletronorte.

In fact, it was only after domestic opposition emerged to lobby for reforms that gradually more effective measures were taken by Brazilian authorities with respect to the Tucuruí and other big dam projects in the Amazon. Local organizing by peoples affected by the Tucuruí Project was initiated in 1981 as a result of the difficult situation they faced from the lack of adequate resettlement and compensation. In September of the following year, 400 of the oustees protested for three days in front of a regional Eletronorte headquarters building demanding lands for lands, houses for houses, villages for villages. The non-responsiveness of project authorities prompted another such demonstration six months later, this time involving nearly 2,000 people. A group of representatives of the affected peoples was then sent to Brasilia to lobby and negotiate with federal authorities. Pressured by Eletrobrás and the Ministry of Mines and Energy, and facing continuing protests in the region, Eletronorte verbally conceded to many of the movement's demands.[37]

While the first field works for the building of the Tucuruí Dam were begun in 1975, it was three years later that planning was initiated to resettle the Parakana Indians. But Brazilian authorities, including FUNAI, failed to implement the proposal to relocate the two Indian villages for another six years. It ultimately took violent clashes between the Parakana and incoming colonists for the federal government to pass a decree in 1985 to create the Parakana Indian Reserve.[38] Surveys of the other mostly small farmer families to be relocated were not completed well into the 1980s, and then only because of the sustained grassroots mobilization. Even after Eletronorte's concessions of 1983 and the project's completion in 1984, problems with resettlement and compensation continued. As in the case of the Itaparica Project and others completed during the 1970s and 1980s, the movement around the Tucuruí Dam continued to mobilize as well as lobby project authorities and government officials to improve conditions of the peoples negatively affected.[39]

Similar patterns of partial reform occurred in the numerous Brazilian big dams constructed from the 1970s through the mid-1980s. But comprehensive cost-benefit analyses including environmental effects were rarely if ever conducted on these projects, because Brazil's laws requiring reviews of environmental impacts were not passed until 1986. Balbina, another well-known big dam project constructed in the Amazon, was also exempt from obtaining an environmental impact assessment, because construction on it began prior to the enactment of the federal regulation. However, two successive heads of Brazil's National Environmental Secretariat, SEMA, opposed the Balbina Project during the 1970s and 1980s, which contributed to delaying its completion.

The Balbina Project also had a devastating effect on the Waimiri-Atroari Indians. The dam's reservoir flooded two villages of this nearly extinct tribal people and displaced one-third of their remaining population. A few domestic nongovernmental organizations, such as the Pastoral Land Commission and rural workers' union, attempted to assist the Waimiri-Atroari.[40] But the absence of mobilization by the Waimiri-Atroari themselves, lack of support from transnationally allied nongovernmental organizations, and still weak domestic institutionalization of globalizing norms about indigenous peoples during this period resulted in virtually no reforms of the project itself. The World Bank, under increasing pressure from the transnational campaign being waged against it during the 1980s (see chapter 6), did decline to directly fund the Balbina Dam on environmental grounds.[41]

Widespread criticism of the human, environmental, and financial costs of the Tucuruí and Balbina Projects was critical to mounting domestic and transnational pressure to substantially reform if not halt further big dam building in the Amazon.[42] More specifically, anti-dam actors quickly organized themselves to oppose construction of Eletronorte's proposal to build fifty dams along the Amazon's Xingu River in the mid-1980s. They focused, in particular, on the massive six dam Altamira-Xingu Hydroelectric Complex, which was to cost between $10 billion and $20 billion, generate 17,000 megawatts of energy, flood thousands of square kilometers of land, and displace 16,000 people from no less than eight local groups. Authorities of Eletronorte and Brazilian government officials petitioned several international development agencies for support to build the Xingu Projects.

But a transnational campaign consisting of high levels of grassroots mobilization, domestic lobbying based on Brazil's new regulations for protecting the environment and indigenous peoples, and international pressure against World Bank funding overwhelmed the projects' proponents. By this time, moreover, the process of re-democratization in Brazil had moved along—the new constitution itself was ratified in 1988—opening greater space for political mobilization by anti-dam actors. The Pro-Indian Commission of Sao Paulo (CPI-SP), a nongovernmental organization dedicated to protecting the rights of Brazil's indigenous peoples, took the lead in the campaign. It circulated scathing critiques of the projects, organized a monitoring program to ensure that authorities followed all of the recently enacted governmental regulations with respect to the building of big dam projects, and initiated links with the Environmental Defense Fund, International Rivers Network, and other transnationally allied nongovernmental organizations to begin lobbying the World Bank.[43]

The plight of the peoples, especially the indigenous groups who were negatively affected by the Tucuruí and Balbina Dams, clearly sparked the organization and mobilization of the peoples who were going to be affected by the Xingu Projects. Chief Paulinho Paiakan of the Kayapo Indians and United States anthropologist Darrel Posey coordinated a five-day meeting in February 1989 of more than 1,000 leaders of 20 Indian tribes, environmental and indigenous rights activists, international rock and media stars, and even politicians to discuss the projects. A chief of the Gavião Indians who had been displaced by the Tucuruí Dam strengthened the opposition of the attendees to the big dams when he recounted the misery faced by his people: "They said they would compensate us, but Eletronorte blocked our claim in court. You can't trust them. They say they are only conducting studies. They told us that. But with each study, they sealed our fate. Little by little, they moved in. Then the dam was built." Tuira, a Kayapo women, summed up the conclusions of the delegates when she spat on Jose Antonio Muniz Lopes, Chief Engineer of Eletronorte, in the midst of his defense of the projects, and told him, "You are a liar. We don't need electricity. Electricity won't give us food. We need rivers to flow freely: our future depends on it. We need our forests to hunt and gather in. We don't want your dam."[44]

The transnational campaign was well under way and beginning to have an impact by the time of the Altamira convention at which resistance to the projects was consolidated. Chief Paiakan, Posey, and another leader of the Kayapo, with the assistance of nongovernmental organizations in Washington, D.C., had already met and discussed their opposition with U.S. Congress-members and World Bank officials. Brazilian officials reacted to the lobbying efforts by charging that the Indians had violated a national security law against foreigners interfering in domestic politics. Domestic and international criticism of the action, however, created further momentum behind the transnational campaign against the Xingu Dams. European indigenous rights and environmental groups, like the Berne Declaration, held a series of demonstrations and extensively lobbied the World Bank not to fund the projects.

The Xingu campaign not only proved that sustained domestic mobilization could be effective in blocking the implementation of big dam projects, it also resulted in the increasing organization of Amazonian groups to defend their interests. Within one month of the Altamira meeting, the World Bank announced that it would not fund the proposed Xingu-Altamira Complex. Reeling from the loss of international support and continued criticism being leveled against it, Eletronorte declared that it was reconsidering Barbaquara, the 11,000-megawatt linchpin project of the scheme. Inspired by the victory but still wary, opponents of the Xingu and other projects allied with the negatively affected peoples from the Tucuruí and Balbina Dams to form the Committee of the People Affected by Dams in the Amazon (CABA). CABA subsequently became a critical player in the formation of Brazil's broader anti-dam movement.[45]

Indeed, an increasingly organized country-wide movement of peoples affected by dams emerged during the transnational campaign against the Xingu Projects. The first national conference of workers affected by dams, coordinated by CRAB and the Central Workers Union (CUT) with the assistance of the Pro-Indian Commission of São Paulo (CPI-SP) and the Pastoral Land Commission (CPT), was held April 19–21, 1989. In preparation for the national conference, various state and regional conferences were organized, including the Altamira meeting, the 4th General Assembly of CRAB, a convention in the Sao Francisco River Valley, and others.

The first national conference provided a forum for dam-affected peoples and their supporters from different areas of Brazil to share experiences and establish a provisional national coordinating body for the various movements. The declaration drafted at the meeting criticized the Brazilian model of development, exemplified by its hydroelectric dam building strategy, as serving the interests of domestic and international capital and excluding working classes, small farmers, the landless, indigenous peoples and *quilombos* from decision making processes.[46] It demanded that Brazilian authorities address the social and environmental problems caused by already constructed big dams, allow local peoples to participate in current and future energy development planning, and stop providing industries with subsidized electricity.[47]

The country-wide movement became progressively more organized during the 1990s and evolved into an increasingly more visible and influential actor in Brazilian politics. The first national congress of peoples affected by dams was held in March of 1991. At the convention, the "Movimento dos Atingidos por Barragens" (MAB—Movement of Dam Affected Peoples) was formally constituted and March 14 identified as the national day of action of dam-affected peoples. MAB established an office with a small permanent staff in São Paulo to administratively coordinate the various activities of regional organizations and to be a central link to transnationally allied nongovernmental organizations and social movements from all over the world. By its second national congress held in December 1993, MAB voted to initiate the first international conference of dam-affected peoples in conjunction with the International Rivers Network, which had established its own office in Brazil by that time.[48] It also subsequently became a central nongovernmental actor involved in creating the World Commission on Dams (see chapter 6).

But the central focus of MAB, and its constituent members, was the campaign against the federal Ministry of Mines and Energy and Eletrobrás's plans to build hundreds of more big dam projects in Brazil—the lion's share slated for the Amazon. In fact, Eletrobrás made this clear with the publication of "Plano 1987/2010" in which it outlined its vision to fully tap the country's hydroelectric potential by 2010, with the continued financial and technical support from foreign donors such as the World Bank.[49] Thus the decline in big dam building in Brazil identified at the beginning of this section has not been predominantly the result of a declining number of sites available to build these projects.

In contrast, the question of financing the dam-building effort proposed in Plano 2010, which was subsequently revised to Plano 2015 by Brazilian authorities, has proven to be a major bottleneck. Brazil's debt crisis of the 1980s clearly reduced the amount of public sector funds available for big dam projects. But the strategy to borrow money from the World Bank and other foreign funders to fill the gap during the 1990s was substantially undercut, as mentioned earlier, by the successful campaigns waged by MAB in conjunction with transnationally allied groups outside of Brazil.

Building big dams became difficult, furthermore, because the costs of these projects spiraled from mounting domestic and transnational opposition as well as the increasing social and environmental requirements linked to the globalizing norms that became more institutionalized domestically in Brazil. Brazilian officials in turn moved towards the privatization and foreign private financing of big dams and the energy sector as a result of the declining availability of public funds, the growing costs of these projects, and the rise of transnationally allied actors and norms promoting these neoliberal strategies.[50]

But Eletrobrás, and Brazil's federal authorities more generally, did not expect the powerful advocacy that MAB would organize against its plans, with its own support from nongovernmental organizations around the world including the International Rivers Network, Environmental Defense Fund, Oxfam, Bread for the World, and others that were active members of the emergent transnational anti-dam network (see chapter 6). And the trends toward the privatization and international private financing of big dam building in Brazil resulted in domestic opponents of these projects forging even more direct links with other nongovernmental organizations and social movements struggling against neo-liberalism in Brazil and around the world as reflected by the World Social Forum. Thus, transnational organizing, the institutionalization of globalizing environmental, human rights, and indigenous peoples' norms, and domestic social mobilization were indeed critical to transformation of big dam building and development in Brazil during the 1980s and 1990s.

INDONESIA: FAILED MOBILIZATION AMID AUTHORITARIANISM

As in India and Brazil, transnationally allied criticism, the domestic institutionalization of globalizing norms, and increasing levels of local opposition have altered the dynamics of development in Indonesia since the 1970s. But the absence of democracy or, more accurately, the persistence of a strong authoritarian regime backed by a powerful military limited the impact of these factors on big dam building in that country. As Indonesian scholar Colin MacAndrews summarized in the mid-1990s:

> Indonesia's authoritarian political system had its origins in the late 1950s when the country's first president, Soekarno, suspended democratic government and started to move toward more centralized control. Under the New Order government of President (Gen.) Soeharto that came to power in 1967, this trend accelerated, and the political system that exists today is characterized by a marked concentration of power at the central government level. While this kind of authoritarian system is not uncommon in many developing countries, what is striking about the Indonesian system is its remarkable durability.[51]

It was not until 1997, thirty years after the inauguration of Soeharto's New Order regime, that a crisis in Indonesia's record of rapid economic development threatened the persistence of this authoritarian political system.

As in the case of Brazil, and perhaps with even more success, Indonesia's military authoritarian regime executed a top-down, technocratic, growth-oriented, and natural resources intensive model to the country's development from the mid-1960s on. After ascending to power in 1967, Soeharto declared that Indonesia would be the "top dam building nation in Southeast Asia and claimed that water projects would enable lesser developed islands to catch up with Java and become more fully integrated into the national economy."[52] This was accomplished with the increasing financial and technical support of bilateral and multilateral donors. The World Bank, for example, had refused to fund the previous Soekarno government's development plans. But under Soeharto's rule, starting with the first loan approved in 1968, the Bank's lending to Indonesia's authoritarian regime reached levels of between $600 million and $700 million by the end of the 1970s.[53] More than twenty big dams and hundreds of small dams were correspondingly built in Indonesia during the 1970s and 1980s, partially fulfilling Soeharto's vision for the country.

Partly as a result of the major environmental problems caused by Indonesia's rapid economic development, but even more as a result of the rapidly spreading global norms, bureaucratic agencies and procedures with respect to environmental protection became increasingly institutionalized in Indonesia from the 1970s on. Motivated by the U.N. Conference on the Human Environment and follow-up activities to it, Soeharto created a Ministry of Environment in 1978 with the mandate to promote "environmentally sound development" defined as the "conscious and planned endeavor to utilize and manage resources wisely in sustainable development to improve the quality of life." Greater emphasis was placed after 1979–80 on environmental management through Indonesia's major policy planning instruments, including the Broad Outlines of State Policy and the Five Year Development Plans.[54]

Despite its relatively marginal status and low level of resources compared with other ministries, Indonesia's national environmental agency had a progressive and active leader in Emil Salim, who served as minister from 1978 to 1993. Salim, who had served as a member of the World Commission on the Environment and Development in the mid-1980s, was critical to the enactment of pollution regulations and environmental impact assessment procedures applicable to all development projects, and actively supported the involvement of domestic environmental nongovernmental organizations in the work of the ministry.[55] Dam projects became progressively subject to environmental oversight by the state, in particular under Law No. 4 of 1982 and Government Regulation No. 29 of 1986, which required environmental impact assessments and the participation of affected communities in development projects.

Nongovernmental organizations working on development, environmental, human rights, and a host of other issues also proliferated and become increasingly organized during the 1980s and 1990s in Indonesia.[56] As James V. Riker notes, "over 7,000 NGOs had registered with the Ministry of Home Affairs by mid-1990, and over 12,000 were estimated to exist by mid-1994. Buyung Nasution estimates that over 1,000 NGOs are actively engaged in development, environmental protection, and human rights activities in Indonesia," in the 1990s.[57] Regional and national networks of nongovernmental organizations such as the forty-member Yogya NGO Forum (Yogyakarta) and the over 500-member Indonesian Environmental Forum (WALHI) were established in the 1970s and 1980s. Frustrated by the inability to have broader impact under Indonesia's authoritarian system and attempting to focus greater attention on the consortium of foreign donors propping up the regime known as the Inter-Governmental Group for Indonesia (IGGI), domestic nongovernmental organizations subsequently forged a transnational network with Asian and Western nongovernmental organizations, ultimately known as the International NGO Forum on Indonesian Development (INFID).[58]

Widespread and increasing grassroots mobilization around dam projects emerged in Indonesia during the 1970s and 1980s. In an analysis of 93 protest campaigns against water projects that took place between 1971 and 1992, George Aditjondro and David Kowalewski found that 52 percent involved dam construction. The protests occurred virtually all across the country, including 16 out of 27 provinces, and the number of campaigns grew in number over time: while only two were found per year in the 1970s, this figure increased to five annually during the following decade and to nine per year by the 1990s. As the authors state, "Indeed, while the number of new large dams in Indonesia increased during the time period, the number of campaigns increased even faster," and, "mobilization of large numbers of citizens typified the campaigns."[59]

But while some of the protest campaigns achieved some of their goals, few were broadly effective. For example, Sundanese villagers opposed a dam project in West Java financed by the United State Agency for International Development in the early 1970s and forced it to be shelved. The demands of Riau villagers for a resettlement review, greater participation in decision making, and higher rates of compensation vis-à-vis the Kotopanjang Dam funded by the Japanese Overseas Economic Cooperation Fund, on the contrary, were only partially met. Despite support from student, human rights, and environmental groups plus transnational lobbying efforts with Diet members in Japan, the dam was completed and many of the protestors remained unsatisfied with the outcome. In general, Aditjondro and Kowalewski found that of the cases of protest campaigns they analyzed, demands for adequate compensation were unsatisfied in 62 percent, for participation were unfulfilled in 60 percent, and to cancel confiscation of property denied in 54 percent. Overall, no concessions were awarded in 45 percent of the protest campaigns and only partial concessions in 21 percent.[60]

The key factor minimizing the relative effectiveness of many of these campaigns was Indonesia's persistent and powerful authoritarian regime, which allowed the government and other dam proponents to circumvent, inhibit, and/or repress activists contesting dams. For example, "Protestors suffered costs ranging from minor intimidation to murder in over one-fifth of the cases" that Aditjondro and Kowalewski examined.[61] The demonstration effect of these and other repressive tactics cannot be underestimated. In fact, memories of the military violence that resulted in the mass murder of more than 500,000 supposed communists in 1965 and that brought Soeharto to power remain vivid for most Indonesians. While the regime's relationship with nongovernmental organizations was not consistent, repeated laws and decrees were passed to regulate, co-opt and intimidate civil society actors. The threat of violent repression was omnipresent.[62]

Moreover, most Indonesian political institutions were highly undemocratic and thus not consistently available to civil society actors to gain access to or leverage decision-making processes. Despite its increasingly open criticism of the regime, the press generally remained heavily censored, even self-censored, as demonstrated by the overnight shutting down of two major Jakarta-based periodicals in 1987 and 1990. Only two non-regime political parties were legally sanctioned, and neither was allowed to sponsor or participate in grassroots activities.[63] Finally, the judicial process was hardly independent from the interference of government officials, particularly the executive branch. As a former Indonesian Supreme Court Chief Justice remarked in 1995, "We often receive requests from the government to cancel or delay the execution of a Supreme Court ruling in the name of development."[64]

Perhaps the best known and most vivid case of a protest campaign that did not achieve its goals largely as a result of Indonesia's authoritarian regime—and in spite of transnationally allied lobbying, the presence of domestic institutions reflecting globalizing environmental norms, as well as significant levels of grassroots opposition—was that of the Kedung Ombo.[65] Investigations and planning of the Jratunseluna River Basin Development Project, involving five big dams including the Kedung Ombo, began as early as 1969. But it was 13 years later before construction of roads and other infrastructure was initiated that the 5,390 households or nearly 25,000 people to be displaced by the reservoir began to be notified. Over the following three years, before actual dam construction had commenced, villagers were often forced to sign papers agreeing to compensation for their losses by project authorities.

The people who were to be negatively affected by the Kedung Ombo Project initially pursued a legal strategy with the assistance of local nongovernmental organizations to secure adequate indemnification for their losses. Although there were laws that mandated the provision of appropriate compensation, the plaintiffs were largely unsuccessful, given the weakness of the judiciary in Indonesia's authoritarian system. In fact, attempts by the people to seek legal redress were often met with intimidation and repression, including threats of arrest and even murder. Moreover, the rights of the people to participate in the Kedung Ombo, because of its environmental impacts under Law 4/1982, were ignored, even though the environmental impact assessment conducted in 1983–84 acknowledged that many opposed the project's resettlement plan.

The affected people initially did not pursue a more confrontational strategy, primarily because they faced government reprisal. But resistance began to grow and the protests became more vociferous, both among the villagers themselves and allied domestic groups, from the mid-1980s on. Student groups in nearby cities, who linked the affected people's campaign to the broader struggle for democratization in Indonesia, were particularly active. As a result of the increasing grassroots mobilization, government officials raised compensation rates marginally but also attempted to split the opposition by offering differential packages to different classes of displaced people. A cycle of opposition, attempted cooptation, and repression characterized the subsequent dynamics of the project.

The Kedung Ombo proved an easy candidate for a transnational campaign because of the presence of mounting domestic opposition to its negative social and environmental effects in addition to the substantial involvement of foreign donors in the project. Over the years, the broader Jratunseluna River Basin Project had been financially supported by the United States Agency for International Development, the Dutch Government, the European Economic Community, the Export-Import Bank of Japan, and the World Bank. The latter two became the primary funders of the Kedung Ombo, especially the Bank, which signed an agreement to lend the Government of Indonesia $156 million of an estimated $283.1 million cost of the project.

The campaign against Kedung Ombo thus fit nicely with the ongoing transnational lobbying effort to reform international development agencies such as the World Bank (see chapter 6). Transnational advocacy clearly altered the dynamics around the Kedung Ombo Project. Beginning with a letter sent to the World Bank by the Indonesian Institute for Legal Aid (YLBHI) enumerating the problems of the dam-affected, the Bank faced increasing pressure to reform the project. The transnational coalition that emerged centered around the recently founded International NGO Forum on Indonesia (at the time INGI) that included domestic nongovernmental organizations such as YLBHI and the Indonesian Environment Forum (WALHI), as well as foreign allies like the Netherlands Organization for International Development Cooperation (NOVIB). As a result of lobbying efforts directly with World Bank officials and indirectly, for example, via the Dutch minister for development, the Bank was compelled to send a resettlement expert to examine the situation. The consultant's report was highly critical of both Bank and Indonesian authorities.

Faced with transnational lobbying and increasing grassroots resistance, the Bank responded with new funds for improving the resettlement process and Indonesian officials authorized novel reforms, including three new resettlement villages nearer to the reservoir where 850 families could relocate. Yet, in January

1989, the sluice gates of the dam were closed, even though 1,800 families still lived in the area. A local Catholic priest, Father Mangoen Wijaya, began an effort to help women, children, and other weaker people affected by the project, but this initiative was forbidden by the governor of Central Java because he saw it as an attempt to support political resistance in the area. Students responded to the events with demonstrations at the dam site while transnationally allied non-governmental organizations sent a 15-point aide-mémoire to the World Bank condemning the human rights violations that were being perpetrated and once again recommended broad reforms.

Despite the combination of transnational pressure via the World Bank on project authorities and domestic social mobilization, the dam-affected were ultimately frustrated by Indonesia's authoritarian political system. In June of 1990, a lawsuit petitioning greater compensation was filed by a number of families in a provincial court. The judge ruled against the plaintiffs stating that they did not have a claim because they had refused the compensation offered to them by the government. Lawyers appealed the decision, eventually all the way to the Supreme Court.

Although the Supreme Court judgment was not announced for almost a year after it was decided, it surprisingly reversed the decisions of the lower courts and awarded the remaining plaintiffs an unprecedented sum of $4.5 million in compensation. But reaction from Soeharto and his top ministers was intensely critical. The President himself met with the Chief Justice of the case, who mysteriously retired two days after the ruling was announced. Under pressure from the executive branch, within four months rather than the usual four year period, a review panel of Supreme Court Justices annulled the judgment in November 1994. One year earlier, the World Bank had disbursed its last funds for the Kedung Ombo Project.

Thus, despite the mounting opposition to the Kedung Ombo and other similar projects in Indonesia from the 1970s onwards, plans to build other big dams were not shelved. On the contrary, construction of three such projects in the Jratunseluna River Basin Development Scheme was approved by Indonesian officials in the mid-1990s, just shortly after the Kedung Ombo sluice gates were closed.

Indeed, according to the most reliable statistics available, big dam building not only did not decrease, as in India or Brazil from the 1970s onward, it increased. Indonesia built one big dam during the 1960s, 10 during the 1970s, 27 during the 1980s, and an estimated 28 during the 1990s.[66] As of 1997, a further 1,231 megawatts of hydropower was under construction and 4,471.8 was in the planning process.[67] Thus, in Indonesia, the availability of sites or funding clearly did not pose a hard constraint on the construction of more big dam projects.

Given the authoritarian institutions in the country, moreover, the effect of transnationally allied lobbying, the domestic institutionalization of globalizing environmental norms and grassroots social mobilization was limited.

The 1997–98 economic crises in Indonesia, however, undermined both the domestic financial capacity to build big dams and opened up greater space for transnationally allied pro-democracy coalitions and movements.[68] As with the case of Brazil during the 1980s, big dam opponents joined the growing ranks of civil society actors struggling to democratize the Indonesian regime. With greater democratization, the expectation was that the dynamics of big dam building and development more broadly would likely be altered more noticeably in that country during the coming years of the 21st century.

SOUTH AFRICA: DEMOCRATIZATION WITH WEAK DOMESTIC SOCIAL MOBILIZATION

A progressively more consolidated transnational anti-dam network, in which the International Rivers Network, Narmada Bachao Andolan, Berne Declaration, Assembly of the Poor, MAB and other critics were central actors emerged from the series of transnational anti-dam campaigns waged during the 1970s and 1980s (see chapter 6). But despite the increasing pervasiveness and sophistication of the transnational network at the international level, big dam projects were rarely halted in countries without the presence of organized and sustained domestic anti-dam opposition. This is the case even in political regimes that have undergone a substantial degree of democratization, such as in South Africa (and Lesotho) although less so during the 1990s.

Authorities and proponents in South Africa built the largest number of big dams in Africa during the 20th century, in a similar fashion to the other cases examined in this chapter. After 61 big dam projects were completed during the 1950s, 104 were built during the 1960s, 125 during the 1970s, and 124 during the 1980s—most of these at the apex of apartheid rule in the country.[69] Nevertheless, as will be demonstrated later in this section, sites for more big dams still exist in that country and throughout the region.

South African big dam building and development are vividly captured by the changing dynamics around these projects in the Orange River Basin, especially through a comparative-historical analysis between the Orange River Development Project (ORDP) of the 1960s and 1970s and the Lesotho Highlands Water Project (LHWP) of the 1980s and 1990s.[70] Proposals for water diversion projects for the Orange River Basin date back to the early 1920s, and preliminary investigations for potential big dams began in the 1940s. The predecessor to the

LHWP, a project to pump water from parts of the Orange River Basin located in Lesotho to the Orange Free State in South Africa, was proposed in the mid-1950s although not approved by the British imperial authorities of the Lesotho protectorate.

Indeed, the push for a grand big dam building scheme in the Orange River Basin only grew with the rise to power of the Nationalist Party (NP) in independent South Africa during the 1950s. The ORDP, which included the flagship Gariep and Van der Kloof big dams, was subsequently initiated by this apartheid-promoting government in the early 1960s. The NP's desires to support the development of white-owned agriculture, stem the increasing outflow of financial capital that was prompted by international criticism of its repressive racist policies, and build a monument and political symbol to apartheid— all prompted the decision to build the ORDP.

The first of the six phases of the quickly assembled multi-purpose ORDP was approved in 1962–63 on the basis of cursory technical studies and economic analysis. Environmental impacts were barely investigated, although some sedimentation and soil studies were conducted. Compensation was not allocated for black and colored farmers, although it was offered to white farmers to be displaced by the Gariep and Van der Kloof big dams. However, despite organized lobbying and even taking the authorities to court several times, even white farmers who owned land were not able to increase the quantity or improve the nature of their packages. Project planning and implementation involved very little information disclosure or public participation.

Big dam building in South Africa proceeded very much along these lines during the 1970s and 1980s. The Gariep and Van der Kloof big dam components of the ORDP were completed in the 1970s along with more than 100 of these projects during that decade and the following as well. Policy proposals for more comprehensive economic analyses, higher prioritization of social and environmental issues, and increased public participation with respect to decision making around the ORDP specifically and big dam projects more generally did emerge in the 1980s, such as in the 1986 Government Report on the Management of Water Resources in the Republic of South Africa.

But actual practices were slow to change. In that same year of 1986, a clandestine treaty was signed between South Africa's apartheid authorities and Lesotho's military government initiating the LHWP in the Orange River Basin. The 1986 LWHP Treaty proposed the construction of five big dams in Lesotho over a thirty-year period and became one of the largest infrastructure projects under construction in the world from the 1980s on. The scheme basically entailed the delivery of water to the dry Gauteng Region of South Africa in exchange for which the Lesotho Government would receive royalties that correspond to between 3 and 5 percent of the country's gross domestic product until the year 2044. According to the treaty, South Africa took responsibility for all the costs of the project, estimated to be $8 billion, provided all the dams were built. The Lesotho Government established the Lesotho Highlands Development Authority (LHDA) to implement it.

Given the presence of the authoritarian regimes in both South Africa and Lesotho and complete absence of grassroots resistance at the time of its inauguration, construction on Phase 1A involving the giant 185-meter Katse Dam—the largest in Sub-Saharan Africa to date—began with little controversy three years later.[71] Several prominent multinational companies won contracts to work on the LHWP starting as early as 1986 including Acres, Spie Batignolles Balfour Beatty, and Impreglio. The governments of Lesotho and South Africa secured a sizeable portion of the financing for Phase 1A from foreign donors, including $110 million from the World Bank, $57 million from the European Development Fund, and $50 million from the African Development Bank. Another $67 million was borrowed from multinational commercial banks such as Dresdener and Credit Lyonnais.[72]

Given the criticism it was facing with respect to other big dam projects around the world, the World Bank attempted to proactively ensure that compensation and environmental policies and practices were elevated to emerging global standards in the LHWP. As part of the treaty, the two countries agreed that, "members of local communities . . . affected by Project related causes shall be enabled to maintain a standard of living not inferior to that obtaining at the time of the first disturbance. . . . ," the same language that was included in the Bank's policies in this area.[73] It is unclear how many people in total would be negatively affected or how much land lost if the entire scheme is ultimately completed, but the figure was estimated as 35,000 people and 7,000 hectares from the first two dam projects.

Environmental concerns were marginalized early on as the initial 1986 feasibility study concluded that the project posed no insurmountable problems in this area, despite the growing focus on these issues in South Africa. But an international "Panel of Environmental Experts" was quickly appointed in 1988 by the Bank to advise project authorities on environmental (and social) mitigation efforts. The creation of a development fund to support income-generating projects for the dam-affected followed in 1991.

Explicit programs to meet these normative standards on resettlement and compensation were not implemented until after local groups increasingly linked to transnationally allied critics began monitoring the project. Indeed, some independent local organizing vis-à-vis the LHWP emerged during the late 1980s under Lesotho's authoritarian regime with the primary purpose of monitoring resettlement and compensation efforts but not for opposing the project. Based on the initial work of the Mennonite Central Committee in 1987, one of the many church groups working in Lesotho, and an ecumenical church leaders conference held the following year, the Highland's Church Action Group (HCAG) was established to assist dam-affected peoples. HCAG remained virtually the sole domestic nongovernmental organization working in the area, yet it faced a series of funding and organizational problems over the 1990s that hindered it from becoming a powerful advocacy group.[74]

By the early 1990s, the domestic, regional and international policy and institutional context had dramatically changed. The ANC had been elected to govern a newly democratic regime in South Africa in 1994 and a transition from authoritarian rule had also occurred in Lesotho. In fact, a relatively successful process of democratization took place in Lesotho in the first half of the 1990s, at which time the Katse Dam was only in its initial stages of construction and was far from a fait accompli. This involved an initial period of political liberalization, followed by a constitutional commission between October 1991 and March 1992 that included 66 open public meetings throughout the country. Churches convened a national conference attended by several political parties, trade unions, and a range of other nongovernmental organizations later in that year to push the transition along. Campaigning for the announced elections had already begun by that time, as soon as the military lifted the ban on political parties in May 1991. After some delays, free and fair parliamentary elections on March 27, 1993, resulted in the election of Lesotho's first civilian government since 1986, the year of the signing of the LHWP treaty.[75]

Although the process of democratization had its fits and starts after the 1993 election, groups in civil society had much greater opportunities to organize and mobilize with decreasing fear of government repression. The courts were becoming progressively more independent of executive interference and the press more open and free. Perhaps more importantly, South Africa also underwent a transition to democracy during the mid-1990s that opened greater space for civil society organizing around big dams in that country. But neither the affected peoples themselves, nor HCAG, nor nongovernmental organizations in South Africa instigated broader grassroots mobilization like those found in other countries around the world, despite the fact that significant discontent with the project existed among all these actors.[76]

During the same period of time that the political regimes in Lesotho and South Africa became more democratic, transnationally allied nongovernmental organizations also increased their involvement in the LHWP. Christian Aid, a transnational nongovernmental organization headquartered in the United Kingdom, began supporting HCAG and other domestic, church-based nongovernmental organizations working with the affected peoples. The weak implementation of the improved resettlement and environmental policies resulted in more pressure on the World Bank, as well as the governments of Lesotho and South Africa, from transnationally allied nongovernmental organizations and HCAG. Christian Aid published a critical report based on an on-site field visit by one of its consultants in January 1994.[77] The Christian Aid report corroborated submissions by the Panel of Environmental Experts which stated that "even with the implementation of the Rural Development Plan, it will not be easy to meet the requirements of the LHWP treaty and order."[78] By 1995, the Environmental Defense Fund and International Rivers Network also began lobbying the World Bank to halt the project in light of its continuing social and environmental problems, and Christian Aid published an even more critical follow-up report in March of the following year with the Christian Council of Lesotho.[79]

But the continued absence of strong domestic social mobilization limited the potential impact of transnational lobbying in altering the dynamics of the Lesotho Highlands Water Project.[80] Sluice gates on the Katse Dam were closed in 1995 and preliminary infrastructure work began on the Mohale Dam in 1996. Outstanding social and environmental issues related to the Katse Dam had still not been resolved. Studies demonstrating that demand-side conservation efforts could postpone construction on the Mohale Dam for up to twenty years by providing water at a lower cost were dismissed by the two governments.

Indeed, despite this evidence and the gradually growing opposition of nongovernmental organizations in South Africa to the project, transnationally allied campaigning to postpone work on Mohale Dam and halt foreign financial support was limited in its effectiveness.[81] In contrast, in 1998, when local public protests against the Lesotho government occurred, the South African government sent in troops to restore order. Dozens of people were killed when the shooting was over, including 17 at the Katse dam. Protection of the LHWP, South Africa's largest investment in the region, was clearly a primary objective of the troops.

Hence, the emergence of a transnational anti-dam network, the progressive institutionalization of globalizing norms on human rights and the environment, and the return to democratic political regimes in South African and Lesotho resulted in only partial reforms in big dam building because of the lack of strong domestic mobilization in either of these countries. To be sure, the lack of collective action around big dams in South Africa in particular is linked to the more general demobilization of civil society that occurred in conjunction with the transition to democracy in the early 1990s.[82] While domestic critics such as the Environmental Monitoring Group have attempted to spur local opposition to big dams, they have remained relatively unsuccessful in organizing wide spread mobilization.

Thus, many big dams continued to be planned and implemented in South Africa along with the continued governmental support for the LHWP. While it is estimated that fewer than 50 big dams were built in South Africa during the 1990s, this figure would probably have been higher if not for the uncertainties of the democratization process and, in particular, the delays caused by the fact that African National Congress (ANC) government after being elected in 1994 insisted on reviewing all the projects that had been approved by apartheid officials in the earlier government. For example, the first ANC Minister of Water Affairs, Kader Asmal, announced that new big dam projects would not be initiated until new water policies and laws were adopted. The subsequent White Paper of 1997 prioritized investment in water conservation programs, although it also stated that "the development of water infrastructure, including the building of more dams" would still be required."[83] Clearly not only sites but also funding from domestic as well as international public and private resources was available to build more big dams, as the Lesotho Water Highlands Project case demonstrates.

CHINA: ABSENCE OF MOBILIZED OPPOSITION
UNDER A DETERMINED DICTATORSHIP

China has been the most prolific dam builder in the history of the world. Only twenty-three big dams were built in the country before 1949, dating all the way back to the first completed in the time of the Qing dynasty (221–207 B.C.E.). But during the last five decades over 20,000 big dams, and more than 80,000 dams in total, have been constructed! As Shui Fu asserts, "The first dam construction boom in the 1950s was a thrilling time. People's communes, the Great Leap Forward, and the manufacture of iron and steel all stimulated the construction of more hydropower projects. Leaders boldly approved projects to accumulate more water for irrigation without knowing whether they were feasible."[84] At the time, Chinese leaders projected that by 1972 hydropower would constitute more than fifty percent of the country's energy supply.[85]

Correspondingly, as part of the Maoist growth-oriented and natural resource-intensive development strategy executed between the 1950s and the 1970s, China built an average of 600 big dams per year.[86] Some Chinese elites did voice concerns about the devastating environmental effects that the unbridled dam construction program was causing. As scientific leader Zhou Enlai stated in 1966, "I fear that we have made a mistake in harnessing and accumulating water and cutting down so much forest cover to make way for more agricultural cultivation. Some mistakes can be remedied in a day or a year, but mistakes in the fields of water conservancy and forestry cannot be reversed for years." Still, nothing could halt the building spree once it was under way, neither the approximately 10 and 40 million people estimated to have been displaced by these projects nor the more than 3,000 dams that had collapsed by the 1980s.[87]

Despite the tremendous magnitude of China's dam building efforts since the 1950s, the country has far from utilized all of its sites or exploited all of its hydropower potential. According to international dam industry projections, China has only developed about fourteen percent of its technically feasible hydropower potential and hydro-capacity was scheduled to double between 1995 and 2010.[88] Building of big dams for flood control, irrigation, and power has remained a high priority for the Chinese Communist Party and resources continue to be allocated to the Ministries of Water Resources and Power to forge ahead with these projects. China's state has also been increasingly able to finance its development initiatives, including big dams, with both international public and private funds. A large percentage of foreign direct investment to the third world went to China during the 1990s, a large portion from overseas Chinese.

Even with the mounting transnational opposition to big dams, and domestic institutionalization of norms on the environment that could have promoted substantial reform or the halting of these projects, China continued on its path with very little change because of the persistence of its powerful authoritarian regime and the weakness of civil society organizing within the country. China, like other third world countries, began to adopt environmental norms primarily in response to international and domestic concern over China's deteriorating environment after the United Nations Conference on the Human Environment in 1972.[89] A National Environmental Protection Agency was established by the mid-1970s, followed by similar departments at the provincial and local levels. The first environmental impact assessment regulation was passed in 1979 and a more comprehensive National Environmental Protection Law was drafted between 1983 and 1989. However, environmental agencies were far from independent and not very powerful relative to other bureaucratic and decision-making structures in China, and environmental legislation was often circumvented.[90]

Without functioning democratic institutions and sustained domestic opposition, the potential for globalizing environmental norms or lobbying by foreign critics to alter the dynamics of big dam building in China was severely limited. The various mechanisms utilized by big dam critics in other countries: relatively independent courts, competing political parties, organizing protests, as well as access to information and a free press, were generally not available in the Chinese political context. While there has been some growth in the number of domestic nongovernmental organizations, particularly in the environmental issue area, they remain relatively weak and still do not fulfill a major advocacy function in the society.[91] Moreover, when grassroots mobilization does occur in China, authorities tended to respond with violent repression, as the well-known events of the pro-democracy movement in Tiananmen Square of 1989 vividly demonstrated.[92]

Perhaps the most telling case that demonstrates the lack of change in China is the Three Gorges Project, on the Yangtze River. As Chinese scientist, journalist, and activist Dai Qing summarizes:

> The Three Gorges Dam will be the largest dam ever built. Its walls of concrete, reaching 185 meters into the air and stretching almost two kilometers across, will create a 600 kilometer-long reservoir. The dam will require technology of unprecedented sophistication and complexity: it will include twenty-six, 680 MW turbines; twin five-stage lock systems, and the world's highest vertical shiplift.
>
> It is also claimed that the project will cause among the greatest environmental and social effects ever: it will flood 30,000 hectares of prime agricultural land in a country where land is the most valuable resource; it will cause the forcible resettlement of upward of 1.9 million people; it will forever destroy countless cultural antiquities and historical sites; and it will further threaten many endangered species, some already facing extinction.[93]

The Three Gorges Project is projected to generate 18,200 megawatts of energy and cost over $20 billion.

Like so many mega-projects (such as India's Narmada Scheme examined earlier in this book), the Three Gorges has had a long and tempestuous history.[94] Dr. Sun Yat-sen first initiated the idea during the nationalist period of the 1920s. Twenty years later, Chinese engineers developed preliminary plans to build dams at the Three Gorges site on the Yangtze River with the assistance of the United States Bureau of Reclamation. In the following decade, after China's Communist Revolution, Soviet technical experts replaced Western advisers in the formulation of the project. The floods of 1954 that killed 30,000 and dislocated one million people sparked greater interest from Mao Zedong in "taming" the Yangtze. Two years later he wrote the following poem, after swimming across the legendary river.

Great Plans are being made;
A Bridge will fly to join the north and south,
A deep chasm will become a thoroughfare;
Walls of stone will stand upstream to the west

To hold back Wushan's clouds and rain,
Til a smooth lake rises in the narrow gorges
The mountain goddess, if she is still there
Will marvel at a world so changed.[95]

At Mao's urging, further studies were conducted to redesign the dam from a flood-control to a multi-purpose mega-project.

But planning was stalled until the 1980s as a result of various bureaucratic and technical controversies surrounding the Three Gorges Project.[96] A five-year agreement, resulting in United States technical assistance to China, was signed in 1981. Two schemes were subsequently formulated, involving dams with heights of 540 feet and 450 feet. By 1985, a United States Three Gorges Working Group consisting of both public (U.S. Bureau of Reclamation, U.S. Army Corps of Engineers, and others) and private entities (Bechtel Corporation, Merrill Lynch Capital Markets, and others) submitted a proposal to the Chinese that reviewed the two options and recommended further social and environmental impact studies. The Working Group also advised that a cost-benefit analysis be conducted and that further financing be sought from bilateral and multilateral agencies like the World Bank, the Asian Development Bank, and the Canadian International Development Agency (CIDA). The next year, CIDA financed a feasibility study to be conducted by a Canadian consortium in conjunction with the World Bank. China's State Planning Commission initiated a parallel study.

Foreign and elite domestic opposition emerged as the initiation of the Three Gorges Project became more probable. Led by the environmental group Probe International of Canada, numerous nongovernmental organizations outside China lobbied their governments and international agencies not to become involved with the project. At the same time, in 1986, members of the Chinese People's Political Consultative Conference (CPPCC, an advisory body to the Communist Party composed mostly of non-communist scholars, bureaucrats, and other professionals) submitted a report to the Central Committee of the Communist Party, entitled "The Three Gorges Project Should Not Go Ahead in the Short Term."[97] The CPPCC's recommendations generated intense debate and resulted in the National People's Congress voting for further examination before proceeding with the project. A 400-member Three Gorges Project Examination Committee was appointed to conduct a final exhaustive feasibility study. Mounting public criticism of the Three Gorges Project in Canada resulted in CIDA and the World Bank commissioning further environmental and resettlement studies two years later.

But by 1989, both the Chinese and CIDA-World Bank feasibility studies concluded that an over 550-foot tall major dam for the Three Gorges would be technically, economically, socially, and environmentally sound and recommended that construction begin in 1992. Nongovernmental organizations from around the world denounced both feasibility studies and ratcheted up lobbying efforts vis-à-vis their governments as well as the World Bank. Some members of the CPPCC immediately criticized the Chinese Government's study as unscientific and undemocratic in a volume of essays edited by Dai Qing entitled *Yangtze! Yangtze!* The *Far Eastern Economic Review* called the CPPCC's efforts "a watershed event in post-1949 Chinese politics as it represented the first use of large-scale public lobbying by intellectuals and public figures to influence the governmental decision-making process."[98] As a result of the criticism, China's National Peoples Congress and State Council postponed a final decision on the project for another five years.

According to sinologist Frederic Moritz, the regime's "loss of face" due to the Three Gorges Project opposition inspired the subsequent student protests at Tiananmen Square.[99] Chinese authorities, however, responded to both the anti-Three Gorges activity and the anti-regime mobilization with the full range of repressive tactics available under their authoritarian regime. The June prodemocracy movement was violently crushed and leaders of the Tiananmen protest were arrested without a trial. Dai Qing was publicly chastised by the official media, and also arrested in July of 1989. *Yangtze! Yangtze!* was banned while the government organized public forums and officially commissioned reports promoted the benefits of the project. Premier Li Peng, a hydraulic engineer by training, in particular, pushed for construction to begin as early as possible. Consequently, in 1992, the Three Gorges Project was sanctioned by China's National People's Conference, although an unusually high one-third of the members abstained or voted against it.

The Tiananmen crackdown and increasing pressure from transnationally allied nongovernmental organizations resulted in, first, a partial Canadian withdrawal from the Three Gorges Project in 1989, followed by the World Bank and ultimately the United States Export-Import Bank during the first half of the 1990s. Despite renewed domestic criticism by Chinese elites such as Dai Qing (who was released after spending eleven months in a maximum security prison in 1990), Chinese authorities nevertheless moved rapidly ahead with the project after 1992. In addition, export credit agencies from Sweden, Switzerland, and even Brazil supported the project with nearly $1.5 billion in credits and guarantees, while transnational private sector banks from the United States, France, and Switzerland also funded it.

China's authoritarian regime prevented domestic anti-dam proponents from effectively accessing domestic decision-making structures as well as from forging sustained coalitions with transnationally allied nongovernmental advocacy organizations. Perhaps most important, however, the lack of domestic social mobilization (noticeably absent from the preceding analysis) limited the ability of opponents to more dramatically alter the trajectory of the Three Gorges Dam.[100] Indeed, without the organization of widespread grassroots resistance by the estimated 1.2 to 1.9 million people to be displaced, it was unlikely that the project would be halted or even substantially reformed.

However, the continuing opposition to the big dam project by a wide array of elites—"intellectuals", scholars, professionals, politicians, and officials—did constitute one of most visible and vocal forces pushing for democratization of the Chinese regime during the 1990s. Calls for greater access to information and open public debate challenged authorities to liberalize politics in the country. As Dai Qing argued in 1997 interview:

> Three Gorges is a metaphor of China's changing society. The politicians who support it have all the characteristics of the old society: authoritarianism, central economic control, and dictatorship of one person. They have no regard for the individual and allow no democratic discussion.
>
> Those opposing the dam represent the new society. They are the majority of intellectuals, who oppose it for scientific, financial, and ecological reasons as well as for human rights and the preservation of cultural relics. So one can actually study China through this case—the whole of contemporary Chinese Affairs.[101]

Thus, as with other third world countries, anti-dam actors may be critical contributors to democratization in China.

If China did transit to some form of even limited democracy, the dynamics of big dam building would likely be more visibly altered in that country. Yet, while the financing of these projects is not likely to pose much of a constraint on further big dam construction, the prolific rate of building between the 1950s and 1970s has made the availability of sites an issue. The likelihood is that with the remaining sites, to the extent that it is at all technically justifiable, mammoth projects like the Three Gorges will be undertaken in China for the foreseeable future, which might generate greater social mobilization.

But the case of the Three Gorges Project also suggests that by the end of the 1990s, emergent transnationally allied critics and the domestic institutionalization of globalizing norms became increasingly powerful factors in development dynamics around the world. Chinese authorities took more and different steps regarding environmental conservation and resettlement than they would have in the past because of the presence of these increasingly powerful actors and authoritative norms. Foreign donors and other powerful actors have likewise had to significantly shift their big-dam building practices vis-à-vis this and other similar projects. Correspondingly, the emergence and consolidation of a transnational network, and its dialectical interaction with globalizing norms and institutions that shape big dam building and development dynamics more generally are the themes of the next and final chapter of this study.

❖

CHAPTER SIX

Dams, Democracy, and Development in Transnational Perspective

Even when transnationally allied critics and opponents of big dams encountered unfavorable domestic-level conditions, by the 1990s they generally had a discernible effect on outcomes because of their expanded capacity for coordinated multilevel mobilization and ability to utilize increasingly authoritative globalizing norms and institutions on the environment, human rights, indigenous peoples, among others. In the case of the Three Gorges Project, for example, the emergent transnational anti-dam network prevented several foreign donors and international development agencies, such as the United States Export-Import Bank and the World Bank, from financially supporting the Chinese state in its determination to build one of largest dams in human history.

Correspondingly in this final chapter, the formation and institutionalization of a transnational network in which member groups share a commitment to contesting big dam projects and arguably promoting more participatory, equitable, and sustainable alternatives is examined. This transnational network gradually emerged during the 1980s and 1990s from the links that had been forged among proliferating domestic and international environmental, human rights, and indigenous peoples nongovernmental organizations, with the growing number of peoples' groups and social movements in developing countries and elsewhere around the world.

Moreover, transnational coalitions and this transnational network contesting big dam projects were built upon, and also contributed to the strengthening of, increasingly organized and overlapping transnational coalitions and networks focused on issues such as the environment, human rights, indigenous peoples, gender justice, anticorruption and good governance, among others.[1] The transnational coalition targeting multilateral development banks that highlighted the destruction caused by large-scale development projects and especially big dams proved particularly critical to the effectiveness of specific transnational anti-dam campaigns and the formation of a broader transnational network during the 1980s and 1990s.[2] This transnational network, moreover, gradually began to advocate for more socially just and sustainable water resources, energy, and river basin development and management options in addition to opposing big dams.

As a direct result of the increased effectiveness of the mounting transnational opposition to big dams interacting with the global spread of norms that together contributed to delegitimating and denaturalizing these projects, a world commission composed of representatives from nongovernmental organizations and social movements, the private sector, states, professional associations, and scientific communities was established in order to formulate and promote the worldwide adoption of novel norms on big dams in the context of sustainable water resources, energy and river basin management for the 21st century. Not only the mandate but also the composition of this commission, in which leaders of four nongovernmental organizations and social movements critical of big dam building were among the twelve commissioners selected to negotiate these norms, was unprecedented.

This type of multi-actor, multisectoral, multilevel "global governance" arrangement was unimaginable for most of the 20th century and provides clear evidence that the international big dam regime had been significantly altered by the late 1990s. Given this institutional shift, the visibly reformed policies and practices of pro-dam actors such as states and multilateral development banks, and the dramatic decline in big dam building across the third world more generally, it is clear that a profound transformation in the transnational political economy of development had occurred in the last quarter of the 20th century. Indeed, the transnational structuration dynamics that conditioned these changes with respect to big dams depicted in this book were both linked to and indicative of similar dynamics that contributed to the transformation in development more broadly.

THE RISE AND IMPACT OF OPPOSITION TO BIG DAMS IN THE WEST

The experiences and effectiveness of dam critics and anti-dam groups in the United States and several countries in Western Europe from the 1950s on were crucial contributors to the formation of transnational coalitions and the gradual emergence of the transnational network contesting big dams during the

1980s and 1990s. In Western Europe, particularly in countries such as Austria and Sweden, most big dams were built between the 1920s and 1960s. In France and the United Kingdom, many more of these projects were initiated after World War II, spurred by the Marshall Plan. These projects were built to generate hydroelectric electricity in countries that had few other energy sources, to supply water to rebuilding and growing cities, and to provide irrigation at subsidized rates to small but extremely vocal agricultural lobbies.[3]

During the 1960s and 1970s, criticism against big dams emerged in many European countries, including Norway and Switzerland, where these projects had historically played an important role.[4] New laws requiring public disclosure of information and mandating environmental impact assessments on big dam and other large projects were both the result of these efforts and contributed to the effectiveness of these efforts. The campaign against Norway's Alta Dam between 1970 and 1981 (in which reforms including improved environmental regulations and the first parliament for the Sami ethnic minority were engendered) and the several successful campaigns during the 1970s and 1980s of the River Savers' Association in Sweden are indicative cases demonstrating these changing dynamics.[5]

The rise of this opposition, in conjunction with the decreasing availability of sites and saturation of demand for the services provided by these projects resulted in a clear decline in big dam building by the 1970s. A well-known campaign, opposing one of the last big dam building schemes in France, demonstrated the dramatically declining likelihood that these projects would be constructed in Western Europe during the 1980s and 1990s. The development program involved the construction of four big dams and hundreds of kilometers of dykes on the Loire River in order to generate irrigation for agriculture, to provide water for cooling processes of nuclear plants, and to control floods. The campaign known as the SOS Loire Vivante within France resulted first in the cancellation of the most symbolic of the big dams, the Serre de la Fare. Subsequently, in a complete policy reversal, the French Government launched a program for flood management, river restoration, and the conservation of biodiversity of the Loire Basin based on the alternative proposals of environmentalists and other big dam critics.[6]

Partially as a result of these successful campaigns and the broader decline of big dam building in most European countries, many European big dam proponents progressively moved their activities to countries where the demand for these projects was still high, anti-dam opposition was muted, and democratic and environmental norms less institutionalized. In response, European nongovernmental organizations, such as the Ecologist, Survival International, Berne Declaration, Urgewald, and the European Rivers Network increasingly focused their energies on halting big dam projects being promoted by public authorities and private actors in Southern Europe, Eastern Europe, the former Soviet Union, and especially across the third world. These indigenous peoples, human rights, and environmental nongovernmental organizations also contributed to the formation of the transnational advocacy network contesting big dams by consciously building links with allies from all over the world.

As in Western Europe, the earliest domestic campaigns around dams in the United States were led primarily by conservationists trying to preserve the natural environment, especially wilderness and other scenic areas. These early environmentalists, fighting to protect the natural beauty of the "American West," were quite effective: for example, they stopped construction of the Echo Park Dam in 1956 and two other big dams proposed for the Grand Canyon in 1967. These struggles around the Grand Canyon projects during the 1950s and 1960s signaled the end of the expansion years of big dam building and played a central role in instigating the broader environmental movement in the United States.[7]

The Grand Canyon campaigns began in reaction to the initiation of the 525-foot-high Echo Park Dam on the Green River, the Colorado River's largest tributary, during a time of mounting environmental activism in the United States. This gigantic project generated immediate opposition from conservationists because it was to inundate Dinosaur National Monument, considered to be one of the country's greatest natural wonders. Howard Sahniser, the executive Secretary of the Wilderness Society, organized a coalition of conservation and recreation groups to lobby the U.S. Congress against the project.

The lobbying effort against the Glen Canyon Dam was energized when executive director David Brower of the increasingly prominent Sierra Club reversed the organization's earlier stand that the monument was not worth saving and threw its weight behind the campaign. Well-known conservationist Wallace Stegner completed a series of books demonstrating the majestic beauty of Dinosaur Monument at the urging of Brower. The books were sent to members of Congress, high-level bureaucrats, and press editors from around the country in a widespread effort to shift official and public opinion against the project.

The financial and economic criticisms of civil society opponents made a significant impact in the campaign. These arguments provided Eastern lawmakers, (already resentful of the large amount of federal money spent on these projects primarily built in Western U.S. states) with evidence to criticize the Echo Park Dam. David Brower's scathing testimony to Congress in 1954, in which he claimed that the Bureau of Reclamation could not be relied upon to have provided accurate figures on the Echo Park Project since it could not "add, subtract, multiply, or divide," was an important turning point. As a result, by

1956, Western politicians strategically gave up the Echo Dam in order to ward off a more general backlash against all large, federally financed water projects. Not surprisingly, later that year, the Appropriations Bill of Congress authorized what would end up being an even larger and arguably more destructive project—the Glen Canyon Dam. After six years of construction, the gates of that dam were closed in 1963.

Conservationists, in particular Brower, so deeply regretted not opposing the Glen Canyon Dam that when, in the same year, the Bureau of Reclamation announced plans for the construction of two more big dams for the Grand Canyon, they mobilized a four-year-long mass campaign against them. The tactics ranged from a public information blitz involving films, articles, advertisements, and even coffee-table books, to lobbying Congress through letters and testimonies backed by strong technical and economic arguments. When Interior Secretary Stewart Udall canceled the Grand Canyon projects in 1967, it represented a huge blow to the pro-dam lobby and a monumental victory for the anti-dam and environmental movements in the United States.

Subsequently, during the 1970s, numerous big dam projects were fought and the anti-dam movement was becoming increasingly organized. Some individual campaigns were less effective than others, such as the ultimately unsuccessful battle against the New Molones Dam in Northern California, but the broader goals of the domestic anti-dam movement were eventually achieved. In 1972, Dr. Brent Blackwater established the Environmental Policy Center (EPC), later to become the Environmental Policy Institute (EPI), to lobby the federal government against large-scale water projects. One year later, the American Rivers Conservation Council (ARCC) was established with an office in Washington, D.C. These two organizations arranged and sponsored the first national "Dam Fighters" Conference in 1976, which subsequently became an annual meeting. Between 1972 and 1983, Blackwelder, EPC/EPI, the ARCC and other anti-dam proponents helped stop more than one hundred big dams from being constructed in the United States.[8]

When the United States Army Corps of Engineers completed the Lower Granite Dam on the Snake River in Washington State in 1975, it was to be one of the last massive river projects to be built in the country. As Robert Devine noted in a historical review of dam building in the United States, "The very success of the dam-building crusade accounts in part for its decline; by 1980 nearly all the nation's good sites—and many dubious sites—had been dammed. But two other factors accounted for most of the decline: public resistance to the enormous cost and pork-barrel smell of many dams, and a developing public understanding of the profound environmental degradation that building dams can cause."[9]

Another turning point occurred in 1987 when the U.S. Congress decided to limit the huge taxpayer subsidies for large water projects and require local beneficiaries to pay 25 percent of their cost. Finally, in 1994, the Commissioner of the U.S. Bureau of Reclamation, Daniel P. Beard, acknowledged that, "we have recognized our traditional approach for solving problems—the construction of dams and associated facilities—is no longer publicly acceptable. We are going to have to get out of the dam building business. Our future lies with improving water resource management and environmental restoration activities, not water project construction."[10]

Moreover, a push for decommissioning big dams in the United States, and around the world, that environmentalists and other critics argued should never have been built in the first place emerged during the 1990s. For example, on November 25, 1997, the Federal Energy Regulatory Commission, a bureaucratic agency that promoted dam projects in the United States for nearly a century, ordered the demolition of the Edwards Dam in Augusta, Maine. This was the first time in that country's history that such a decision was taken with respect to a hydroelectric dam that its owner wanted to continue operating. As Blaine Harden, a reporter for the *Washington Post* wrote, "The federal order is likely to ratchet up pressure across the country, particularly in the salmon-depleted rivers of the Pacific Northwest, for removal of dams that are far bigger and produce vastly more power than Edwards and where the case for demolition is considerably less clear-cut."[11]

Two big dams, the first 210 feet high and the other 100 feet high, on the Elwha River in the northwestern State of Washington increasingly became important test cases for decommissioning in the United States. "Tearing out of the dams has long been a dream of greens, amateur fisherman, and Indian tribes who once fished the river."[12] One of the project's licenses came up for renewal in the early 1980s and others never had a license. During the process of re-licensing, opponents convinced the National Park Service to call for their demolition in 1992. Congress supported the idea, partially because the dams were located in historic Olympic National Park. Two years later, the Secretary of the Interior, Bruce Babbitt, agreed to look into the issue and, partially as a result of his efforts, President Bill Clinton's 1998 budget included $25 million to buy the dams and begin demolition. Another $83.3 million was slated over the next five years to complete the work.

However, the Elwha dams are but two dams in the region. Eighteen dams are located on the Columbia River and its principal tributary, the Snake. These projects combined produce one-third of the total hydropower in the United States. These big dams continued to be defended by a powerful pro-dam lobby including electricity-needy aluminum companies, irrigation-dependent farmers, and large agro-businesses. Such interests saw the fight over the Elwha dams as symbolic, even though the projects produced only 28.6 megawatts of electricity compared to 7,000 megawatts generated by the gigantic Grand Coulee. The pro-dam actors thus organized a counter-lobbying "Save Our Dams" effort against the anti-Elwha coalition.

A campaign by environmentalists (led once again by the Sierra Club) to drain Lake Powell, the massive reservoir created by none other than the Glen Canyon Dam, was also revived during the 1990s. As David Brower, the executive director of the Sierra Club in the 1950s, stated, "I've been kicking myself for 40 years" for letting the Glen Canyon Dam be built in exchange for the scrapping of the Echo Park Project.[13] Lake Powell inundated several thousand Native American Ruins, brought a number of fish and wildlife near extinction, and arguably destroyed some of the most beautiful riverine sites in the United States. Opponents to the draining included electric power companies and their six million customers in the western region of the country, boaters and fly-fishers, and the nearly 8,000 people who live in the nearby town of Page—originally built for dam workers and their families—that depend on the revenues generated by Lake Powell's tourism. Still, the momentum behind perhaps the largest ever restoration campaign in United States environmental history continued to grow as the 20th century came to a close.

No matter what happens in these and other decommissioning cases, the big dam building era in the United States seems to be over. But the counter-mobilization by big dam proponents against decommissioning in the United States was not an isolated event. This suggests that the dramatic changes in big dam building dynamics during the second half of the 20th century and especially from the 1970s partly depended on the passive state of defenders of big dams. Indeed, around the world from the local to the international levels, pro-dam actors became increasingly and more self-consciously organized and mobilized to counter-act the decreased ability to plan, build and operate these projects during the 1990s. If this trends continued, it is possible that the unanticipated changes in big dam building documented and explained in this book may be reversed.

THE EMERGENCE OF THE TRANSNATIONAL ANTI-DAM NETWORK

As big dam building declined in the West, these projects were increasingly being formulated and implemented in the rest of the world. In fact, by the 1980s more than two-thirds of new big dam starts were located in Africa, Asia, and Latin America. Independent local and domestic resistance had already emerged before this shift and grew even more in reaction to the upsurge of big dam building in their countries. But these contentious efforts in developing countries were generally less effective in altering outcomes than in Western Europe or the United States during the 1950s and 1960s (see chapters 2–5).

Part of the explanation for the comparatively low impact of solely domestic campaigns across the third world was ostensibly the demand for the benefits produced by big dams. But the existence of centralized authoritarian and often repressive regimes in most of these countries greatly reduced the opportunity structures for political mobilization as well. Moreover, big dam building was supported by a range of domestically powerful and transnationally allied proponents that were often beyond the reach of these locally based critics. Finally, the hegemonic global norms underpinning visions of development around the world—progress as economic growth, electricity consumption as modernization, nature as technologically controllable, etc.—further legitimated and naturalized the building of big dams.

Opponents soon began to jointly organize structures of solidarity and craft linked campaigns as they realized that these projects had migrated to parts of the world where the relations of power and meaning were even more skewed toward big dam proponents and that these proponents were themselves transnationally connected. As one critic of U.S. big dam building efforts and an early organizer of the transnational anti-dam network recounted,

> As we were becoming successful in stopping dams in the U.S., we saw the same obsolete big dam technology being exported to the third world, disregarding the devastating ecologic damage and huge economic costs that the U.S. had incurred in its dam building boom in the '50s and '60s. Yet at the same time we saw successful examples of citizens groups' resistance in the early 1980s, most notably the cancellation of dams on the Franklin River in Tasmania, the Silent Valley in India, and the Nam Chaom Dam in Thailand. Around the world big dams were being promoted to gullible and corrupt politicians as a quick road to economic development by an international syndicate of self-serving interests: international consultants, construction firms, and development bureaucrats. . . . We believed that by cooperating with and coordinating with nongovernmental organizations fighting dams in international policy arenas we could counter the influence of these international interests.[14]

The campaign against India's Silent Valley Project examined in chapter 2 was one of the first examples of a successful transnational coalition opposing a big dam in the third world.

While the process began somewhat earlier historically, there are remarkable parallels between the transnationalization of advocacy around big dams in the

United States with those in Western Europe. During the late 1970s, Dr. Philip B. Williams, a hydrologist living in San Francisco who had for years supported environmentalists campaigning against large water projects in California, began conducting research on the social and environmental effects of big dams worldwide and, in particular, on their safety records and the structure of the global dam industry. In 1982, after the early victories in the transnational campaigns against India's Silent Valley Project and Thailand's Nam Choam Project, Williams urged big dam-fighter Brent Blackwelder of the Environmental Policy Institute to organize a session on big dam building around the world to be held in conjunction with the annual United States dam fighters' conference in Washington, D.C.

At that time, campaigns contesting big dams were further given a boost with the initiation of a transnational multilateral development banks campaign. Bruce Rich, an attorney with the then Environmental Defense Fund (EDF)—another important environmental nongovernmental organization headquartered in Washington, D.C.—who had been investigating the negative impacts of large infrastructure and other mega-development projects like big dams funded by the World Bank and other multilateral agencies, attended the U.S. dam fighters' meeting at which stopping the building of these projects around the world was discussed.[15] The next year, in June of 1983, Rich of EDF, Blackwelder of the Environmental Policy Institute, and Barbara Bramble of the National Wildlife Federation gave the first ever testimony to Congress on the destructive social and environmental impacts of World Bank supported, large-scale infrastructure projects, many of the worst being big dams, as part of the newly initiated multilateral development banks' campaign.[16]

The multilateral banks campaign coordinated by the EDF, Friends of Earth, Oxfam, Survival International, Cultural Survival, Urgewald, and other nongovernmental advocacy organizations gathered momentum and was increasingly effective in reforming the World Bank and other development donor agencies during the 1980s and 1990s (see next section). But transnational coalitions such as those that formed to contest projects such as Polynoreste in Brazil, the Kedung Ombo Dam in Indonesia, and the Sardar Sarovar Dam in India were critical in broadening what was initially conceived as primarily an environmental campaign. As Paul Nelson suggests, "The agenda of the campaign was initially concerned with protecting river basins, preserving tropical forests and biodiversity, and promoting demand reduction and efficiency in energy lending. However, three related issues have come to equal, and sometimes eclipse, these conservationist themes: involuntary resettlement of communities from dam projects, protection of indigenous peoples' lands, and accountability and transparency at the World Bank."[17]

Overlapping with the initiation of the multilateral development banks campaign, the 1984 publication and worldwide distribution of the first volume of Edward Goldsmith and Nicholas Hildyard's series, *The Social and Environmental Effects of Large Dams,* further launched the transnational anti-dam network. The writings of Philip Williams, Brent Blackwelder, Bruce Rich, and other activists from the United States were among those in the book. As Blackwelder stated in his forward, "The staggering array of problems created by large-scale water development is so alarming and widespread that an international network has been established to halt the destruction and to propose sensible alternatives. Goldsmith and Hildyard's book stands as a landmark in providing the most comprehensive information and analysis to date on the tragic impacts of superdams."[18]

The study was the first to systematically assemble the main arguments against big dams for a transnational audience and to insist that the problems caused were largely inherent to the technology. Social impacts, particularly in terms of the forced displacement of peoples, were fully integrated into the ecological critique of these projects. The analysis also challenged the conventional wisdom that big dams were the most financially and economically viable means by which to generate irrigation and power or control floods.

Volume one drew immediate and widespread international attention because the authors, Goldsmith and Hildyard, were also editors of the well-known environmental journal, *The Ecologist. The Ecologist* had already become a vehicle for the publication of several critiques of big dam projects that had been built, were under construction, or being planned in numerous countries around the world. A second volume of *The Social and Environmental Effects of Large Dams* published in 1986, this time edited by Hildyard and Goldsmith, further fostered the formation and expansion of the transnational network. It gathered together case studies of big dam projects and struggles written by researchers and activists from around the world, a large portion of which focused on the third world.[19] That same year, Stephen Schwartzman published his Sierra Club–sponsored critique of large development projects financed by the World Bank, including numerous big dams, entitled *Bankrolling Disasters.*[20]

The increasingly linked local, national, and international efforts contesting big dams around the world, the multilateral banks campaign focusing on large-scale infrastructure projects, and the publication of the Goldsmith and Hildyard series—all contributed to the establishment of the transnational nongovernmental organization eventually known as the International Rivers Network (IRN). Encouraged by the growing number of campaigns and linkages among groups around the world, Dr. Philip Williams motivated a group of volunteer environmentalists in California to start a bi-monthly "International Dams Newsletter" in late 1985. The newsletter was intended "to help citizen's organizations that are working to change policies on large dam construction throughout the world," and, "as one channel of communication for activists in all parts of the world, so that lessons learned in one place can be put to use in another."[21] The first issues included overviews and critiques of the Three Gorges, Bakun, Sardar Sarovar, and other projects that were considered big-dam "hot spots." With support from the San Francisco-based Tides Foundation and the United States domestically focused nongovernmental organization Friends of the River, 1500 copies were soon being distributed to individuals and organizations in 56 different countries.

Feedback from readers and activists from around the world, as well as a series of internal discussions, resulted in the creation of a more formalized organizational structure to increase the potential impact of the "International Dams Newsletter" volunteer group. In mid-1987, the group formally became the International Rivers Network (IRN) and the newsletter was renamed "World Rivers Review."[22] A conference on the undelivered benefits, social problems, and environmental costs of big dams was organized by IRN in San Francisco for 1988, at which 63 big dam critics from 23 countries met. The participants drew up the "San Francisco Declaration," demanding an independent assessment of big dam projects and a moratorium on all big dam building not having the participation of project-affected persons, free access to project information, environmental impact assessments, comprehensive resettlement plans, and full cost/benefit analyses. They also endorsed a "watershed management" declaration recommending numerous alternatives to big dams.[23]

This San Francisco Declaration, and the subsequent Manibeli and Curitiba Declarations later in this chapter, articulated the evolving sets of shared norms that tied transnationally allied big dam critics together. These and other tactics and tools utilized the increasing global spread of norms on indigenous peoples, human rights, and the environment, and applied them directly to the issue of big dam projects, river basin management, and the provision of water and power services. The development of these pro-sustainable development norms contributed to the strengthening of the transnational advocacy network contesting big dams by providing a common discourse that facilitated communication and joint action among its constituent actors. In addition, these norms were used strategically by big dam critics, first as guidelines for appropriate policy prescriptions and subsequently as means for holding pro-dam actors accountable after these norms were adopted into the procedures and structures of international organizations, states, and multinational corporations.

The deepening connections among allied groups within the transnational network resulted in critical changes in the set of goals and self-understandings they shared, not to mention the range of strategies they learned to employ. By 1989, according to Philip Williams,

> IRN had evolved into a structured organization and our vision had expanded and changed. While our analysis of the problem remained the same, we now understood that the destruction of rivers was as much a social and human rights issue as environmental. We started to see the importance of dams as centerpieces of an inappropriate development ideology and realized we could use dam fights as important weapons in a larger war against institutions like the World Bank or against dictatorial governments.
>
> Our vision of how IRN could operate most effectively in these bigger arenas also changed. As we learned more about the internal dynamics of the international environmental movement, we realized the impracticability of coordinating a formal international structure for IRN, and instead recast ourselves as a U.S.-based organization that derived our legitimacy from the respect of the individual groups we worked with all over the world. We saw ourselves as one part of a larger citizen-based movement working on these issues.[24]

IRN quickly became a leading nongovernmental organization in the expanding transnational network contesting big dams, as well as a member of the ongoing multilateral banks campaign. The organization began publishing another newsletter, *Bankcheck Quarterly,* that contributed to the eventual organization of the Fifty Years Is Enough campaign targeted specifically against the World Bank. IRN also established an office in Brazil and substantially deepened its links with nongovernmental organizations and social movements all over the world.

The transnational network thus grew, became increasingly better connected and more sophisticated, especially from the mid-1980s onward. Numerous advocacy organizations such as Probe International in Canada, the European Rivers Network (ERN) in France, the Association for International Water and Forest Studies (FIVAS) in Norway, Friends of the Earth in Japan, Both Ends in the Netherlands, the Berne Declaration in Switzerland, Urgewald in Germany, AidWatch in Australia, Christian Aid, Oxfam, and Survival International in the United Kingdom, IRN, EDF, Cultural Survival, and the Lawyers Committee for Human Rights in the United States, the Narmada Bachao Andolan in India, the Movimento dos Antigidos por Barragens in Brazil, the Assembly of the Poor in Thailand, and countless others allied themselves together to reform and halt big dam building around the world.

By the 1990s, virtually any big dam being built or proposed in the world became a potential target of a campaign initiated or at least supported by members of the transnational anti-dam network, especially if the project was funded by foreign donors, if transnational companies were involved, and/or if affected people at the grassroots within the country were organized to oppose it. The third volume of *The Social and Environmental Effects of Large Dams,* published in 1992, attempted to review the enormous expansion of critical materials that had emerged as the transnational network grew and became increasingly institutionalized during the 1980s.[25]

Big dam proponents recognized and bemoaned the transnational network's growth and increasing effectiveness. For example, in 1992, the then International Commission on Large Dams' President Wolfgang Pircher warned that the big-dam industry faced "a serious general counter-movement that has already succeeded in reducing the prestige of dam engineering in the public eye, and it is starting to make work difficult for our profession."[26] As the previous chapters demonstrated, increasing numbers of big dam projects were halted during the 1980s, and the policies and practices of more and more states were being reformed. Gradually, leading multinational corporations such as Asea Brown Boveri, Lahmeyer and Bechtel also began to alter the ways they conducted business with respect to big dam building.[27] Indeed with the growing push for greater business involvement in big dam projects during the 1990s, members of the transnational anti-dam network increasingly focused more of their attention directly on these companies.

As the next section will further show, transnational campaigns had a dramatic impact on the organization and functioning of multilateral development agencies such as the World Bank from the 1970s on. Thus, by the 1990s, the norms, principles, rules, and decision-making structures that constituted the extant international big dam regime were discernibly being altered. These changes were shaped by and greatly contributed to the broader transformation in the transnational political economy of development that was also occurring during this time.

TRANSNATIONALLY ALLIED ADVOCACY AGAINST THE WORLD BANK[28]

The campaigns against World Bank-funded big dams have been among the most effective means through which development dynamics have been altered by various transnational coalitions and networks. Moreover, the various campaigns waged during the 1980s and 1990s against World Bank–supported big dams from Thailand's Nam Choam to Brazil's Xingu River to India's Narmada Projects contributed to the formation and institutionalization of the transnational anti dam network.

These transnational campaigns played a leading role in compelling the World Bank to reduce its involvement in big dam projects. But they also motivated the Bank to adopt various new policies, in particular those on resettlement, environmental assessment, indigenous people, and information disclosure. The World Bank's adoption and at least partial implementation of these novel criteria and guidelines signaled to other powerful actors, such as states and multinational companies, that a shift in the extant norms of big dam building and the broader international big dam regime was under way.

Since its founding, as explained in chapter 1, the World Bank has been a powerful player in the transnational political economy of big dam building and development. Correspondingly, the Bank has been a cornerstone in the international big dam regime, particularly with respect to promoting the spread of these projects across the third world. The Bank's first loan to fifteen third world countries was for big dam projects: Chile, Brazil, Mexico, and El Salvador (1949); Iraq (1950); Lebanon (1955); the Philippines (1957); Malaysia (1958); Ghana and Uganda (1961); Swaziland (1963); Bolivia (1964); Syria (1974); Vietnam (1978); and Lesotho (1986). The World Bank's largest borrower through 1993, India, had been lent $8.38 billion for the construction of 104 big dams, far more than any other country in this category as well. World Bank technical support was often critical to the initiation and management of big dam projects while Bank-arranged co-financing generally increased the overall funds available for big dam projects between 50 and 70 percent. The Bank also greatly contributed to the creation of numerous big dam building bureaucratic agencies in the third world, such as Thailand's Electricity Generating Authority and Colombia's Interconexión Eléctrica SA.[29]

Despite the importance of these projects in its portfolio, the World Bank had not conducted a single, comprehensive post-evaluation of the big dams it had funded until the 1990s. It would not have done so at that time had it not been for the increasing pressure it faced from transnationally allied big dam critics from the 1970s on. Nor had the Bank's experience with big dam building produced any significant policy changes with respect to indigenous peoples, resettlement, or the environment.

A few individuals within the Bank, notably Michael Cernea and Robert Goodland, began to become aware of the social and environmental problems around the mid-1970s, but were able to independently affect very little reform. Thayer Scudder, a leading development anthropologist and World Bank consultant, who investigated the social effects of big dam projects for almost forty years, acknowledged that, "During the 1950s and 1960s, engineers, academics, and development practitioners had no clue of the tremendous social and environmental effects of large river basin—big dam projects. While incremental learning began to produce some change at the World Bank by the late 1970s, without environmental and human rights activists and movements, the broader reforms would never have occurred."[30]

Indeed, the World Bank had begun to adopt environmental norms, at least formally and superficially, as early as 1970. Economist Robert Wade states in his authoritative review of the Bank's changing environmental policy from 1970 on that, "As the recognized lead agency in the work on development issues, the Bank could not remain oblivious to this rising tide of concern, particularly because many in the environmental movement were saying that economic growth should be stopped, an idea fundamentally opposed to the Bank's mission. In addition, the Bank had to consider what position it would take at the United Nations Conference on the Human Environment scheduled for 1972 in Stockholm."[31]

The World Bank's President, Robert McNamara, who had been convinced of the importance of environmental issues by prominent conservationists, such as Prince Bernhard of the Netherlands, responded by establishing the post of environmental advisor. The Bank, through its first environmental advisor James Lee, senior Bank official Mahbub ul Haq, who convinced developing country governments worried about their sovereignty that they should participate, and McNamara who delivered the keynote address, played a central role at the Stockholm Conference. Senior management thus attempted to give the Bank a reputation at the time as an international leader addressing environmental issues as they related to development dynamics.[32]

But newly adopted environmental policies and hired specialists played a marginal role in Bank activities throughout much of the 1970s and early 1980s. The Bank's Office of Environmental Affairs, created after Stockholm, remained small during this period: by 1983, the office had only three regular specialists out of a total World Bank professional staff of nearly 2,800. In addition, "the Office of Environmental Affairs was usually involved in project design only at a late stage of the project cycle, at the eleventh hour, too late to modify design." Most Bank staff ignored environmental factors and continued working as if no new requirements existed. When environmental issues were emphasized, the focus was generally on "disease prevention, occupational health and safety inside large factories, and the reduction of air and water pollution from large factories." Such concerns largely neglected the types of human-ecologic, natural resource and ecosystemic issues that were the key planks of the multilateral development banks' campaign of the 1980s and 1990s.[33]

It was only with mounting transnationally allied contestation against big dams around the world, linked to the emergence and increasing effectiveness of the multilateral banks campaign, that substantial change in the World Bank's environmental policies and practices began to occur and novel reforms in other issues areas were initiated. As proposed elsewhere, one mechanism by which reforms were conditioned involved the empowerment of sympathetic Bank staff from the increasing effectiveness of pressure being leveled by critics from the outside.[34] For example, the opposition by the Bontoc and Kalinga peoples against the World Bank funded Chico Dam Project from 1974 on compelled Bank President McNamara to state that "no funding of projects would take place in the face of continued opposition from the people."[35] The Chico Dam and other struggles such as that against the Sobrahindo Dam in Brazil also strengthened the position of World Bank sociologist Michael Cernea to write and have adopted in 1980, with the help of former Food and Agricultural Organization (FAO) resettlement specialist David Butcher, Operational Manual Statement (OMS) No. 2.33, entitled "Social Issues Associated with Involuntary Resettlement in Bank-Financed Projects."[36] These emergent norms on appropriate practices with respect to the resettlement of people displaced by World Bank development projects were a critical institutional breakthrough.

The campaign against the Chico Dam in particular also contributed to increasing attention on the detrimental effects of Bank funded large-scale projects on indigenous and tribal populations. The Bank's subsequent involvement in Brazil's Polynoreste Project, an infrastructure-based colonization program that severely threatened indigenous groups in the Amazon, catalyzed the establishment of principles "to follow in situations where projects funded threatened to infringe the rights of residual ethnic minorities." Moreover, as Andrew Gray, former Coordinator of the nongovernmental International Working Group on Indigenous Affairs, writes: "For about six months before the policy guidelines appeared there were meetings of NGOs and experts concerned with the rights of indigenous peoples, such as the Boston-based NGO Cultural Survival and the U.S. chapter of Survival International. The consultation took place largely with members of the Bank's in-house 'sociology group,' which since 1977 had participated in seminars on questions of social and environmental interest to the Bank."[37]

Big dam critics clearly helped shape World Bank environmental advisor Robert Goodland's report, "Economic Development and Tribal Peoples: Human Ecologic Considerations," and subsequently the Bank's OMS 2.34, "Tribal People in Bank-financed Projects," produced in 1981 and 1982 respectively. As James Lee, the Bank's first head environmentalist accepted, "There were a number of outside groups who were quite vociferous . . . groups like Amnesty International, the Harvard group Cultural Survival . . . and others. They were quick to chastise us and rightly so. . . . And so . . . my office moved out in front on this. . . ."[38] As with the earlier OMS on resettlement, the OMS on tribal peoples established a novel set of policy norms at the Bank for which transnationally allied nongovernmental organizations and social movements had advocated from the early 1980s on.

Initiated by the National Resources Defense Council, National Wildlife Federation, and Environmental Defense Fund in 1983, the basic strategy of the multilateral development banks campaign was to publicize a small number of destructive large-scale projects and use them as levers on member governments to pressure the Bank to institute reforms. Over the following four years, "more than twenty hearings on the environmental and social performance of multilateral banks were held before six subcommittees of the U.S. Congress," and similar events were held across Europe and in Japan.[39] These lobbying tactics effectively conditioned further reforms: in 1984, the World Bank introduced OMS 2.36, entitled "Environmental Aspects of Bank Work" which stated that the Bank would "endeavor to ensure that . . . each project affecting renewable natural resources does not exceed the regenerative capacities of the environment," and that the "Bank would not finance any projects that . . . would severely harm or create irreversible environmental deterioration" nor "displace people or seriously disadvantage certain vulnerable groups without undertaking mitigatory measures. . . ."[40]

The transnationally allied groups in turn effectively utilized the Bank's new guidelines and public commitments to hold them accountable to norms on the environment, human rights, and indigenous peoples. When the Bank made proclamations that it, "with its Office of Environmental Affairs, is the oldest, largest, and most experienced institutions dealing with these issues," campaigners pointed to the Bank's failure to address these issues with respect to the Philippine's Chico Dam Project, among others.[41] Soon thereafter, the Bank suspended its loan disbursements for Brazil's Polynoreste Project in 1985 on the basis of a transnational campaign that identified that project's destructive consequences for the environment and indigenous peoples.

By that time, big dams had begun to take center-stage in the multilateral banks campaign, based on such visible cases as Indonesia's Kedung Ombo, the Ruizizi II Regional Hydroelectric Project in Zaire, Brazil's Plano 2010, and India's Narmada Projects. The campaign, and transnationally allied big dam critics, began to have an impact in 1986 when the United States executive director voted against the World Bank's proposed Brazil Power Sector Loan that was to help finance the construction of 136 dams, the first time any of the Bank's member governments had voted against a loan on social and environmental grounds. Subsequently, transnational lobbying in which two Kayapo tribal leaders and ethno-botanist Darrell Posey traveled to the United States to protest the dams to be built by Brazil with the Power Sector Loan led to a suspension of the second $500 million *tranche* from being dispersed by the Bank (see chapter 5).

Responding to the mounting political pressure, World Bank President Barbar Conable announced on May 5, 1987, that he was "creating a top-level Environment Department to help set the direction of Bank policy, planning and research work" that would "take the lead in developing strategies to integrate environmental considerations into our overall lending and policy activities."[42] One month later, a five-year review of the 1982 OMS 2.34 on "Tribal People in Bank-Financed Projects" conducted by the Bank's Office of Environmental and Scientific Affairs was released. It stated that while the Bank's activities related to indigenous peoples had slightly improved, "there was a general tendency among Bank staff to underestimate the unique social, cultural, and environmental problems that both tribal and indigenous or semi-tribal populations face in the process of development."[43] Senior Bank management renewed the pledge to prioritize issues of environment, resettlement, and tribal/indigenous peoples in project lending. But critics were still not satisfied.

By nearly unanimous agreement, and corroborated by substantial evidence, the transnationally allied opposition to India's Narmada Projects during the 1980s and 1990s produced the most visible change in the policies and practices of the World Bank. The Bank's Operation's Evaluation Department acknowledged this fact in 1995: "The Narmada Projects have had far-reaching influence on the Bank's understanding of the difficulties of achieving lasting development, on its approaches to portfolio management, and on its openness to dialogue on policies and projects. Several of the implications of the Narmada experience resonated with recommendations made by the Bank's Portfolio Management Task Force . . . and have been incorporated into the 'Next Steps' action plan that the Bank is now implementing to improve the management of its portfolio."[44]

The reforms at the World Bank prompted by the conflicts over the Narmada Projects included novel initiatives in a range of areas, from resettlement and environment to procedural mechanisms, that would ostensibly increase the transparency and accountability of Bank activities. For example, the criticism of the high social costs related to the displacement of peoples caused by Bank-funded projects, especially big dams, in the multilateral development banks campaign prompted a 1985 Bank portfolio review of resettlement practices with hydro and agriculture projects over the previous six year period. The portfolio review's conclusion that there was only marginal improvement in Bank compliance with the 1980 OMS on resettlement contributed a significant 1986 policy revision. However, as political scientist Jonathan Fox states, "the 1986 policy revisions led to few improvements (in practice) until after 1992 when awareness of the issue increased due to the Narmada debacle."[45] In fact, resettlement had been a central issue in the transnational contestation over the Narmada Projects from the early 1980s on, as examined in chapter 3.

The second transnational campaign against the Sardar Sarovar Project, as discussed in chapter 4, subsequently compelled the World Bank to constitute the first ever independent review of a Bank-supported project under implementation. The Morse Report concluded that the Bank had not complied with its own resettlement norms and policies, and that it should take a step back from the project because adequate resettlement was not possible under the prevailing conditions in the Narmada River Valley. The conclusions took Bank management by surprise and it responded by sanctioning a comprehensive review of resettlement problems throughout its portfolio. According to Michael Cernea, who proposed the idea, the Bank's managing director, Ernst Stern, "agreed to a new review because it was part of the formal report to the Board and the public answer to the Morse Commission." The Bank was eager to find out if there were "other Narmadas hidden in the portfolio."[46] The Resettlement Review, which was published in 1994, accepted that progress in the Bank's resettlement efforts were often "a consequence of public opinion demands, of resistance to displacement by affected people, and of strong advocacy by many NGOs."[47]

Environmental practices at the Bank were also shaped by the transnational campaigns around the Narmada Projects. As demonstrated in chapters 3 and 4, the Bank's failure to comply with its own 1984 OMS 2.36 on environmental issues and willingness to fund the Sardar Sarovar Project, even though it had not been cleared by India's Environment Department, was successfully exploited by the transnational anti-Narmada coalition. The ability of critics to spur the Bank's process of internal restructuring and policy revision initiated in 1987 was greatly enhanced after links between the then Environmental Defense Fund, Survival International, and the emergent Narmada Bachao Andolan were established in 1986.

The upshot at the Bank was both organizational change, including the appointment of regional environmental directors in 1987, and policy reform, most notably the 1989 Operational Directive 4.00 on Environmental Policy for Dam and Reservoir Projects. The latter mandated that a panel of independent and internationally recognized specialists was to be established to advise on the environmental (and social) aspects of any World Bank–supported dam with a height greater than 10m. This was precisely the policy that led to the appointment of a panel of environmental experts on the Lesotho Highlands Water Project in 1988 (see chapter 5).

But critics, empowered by the effectiveness of the transnational anti-Narmada campaign and angered because they had not been consulted during the drafting of Operational Directive 4.00, quickly compelled a reformulation of the document. The revised policy, Operational Directive 4.01, was completed by October of 1991 with discernible revisions. It stated that, "The purpose of EA (environmental assessment) is to ensure that the development options under consideration are environmentally sound and sustainable." The directive strictly mandated information disclosure and consultation with nongovernmental organizations. It also required borrowing countries to release environmental assessments to the executive directors of the Bank. Moreover, as Robert Wade writes,

> around 1992 and 1993 the more comprehensive ideas of the 'environmental management' paradigm began to take hold at senior management and operational levels. The conversion came partly from love and partly from fear. Narmada was the fear factor. By the early 1990s, staff throughout the Bank were aware of the NGOs' anti-Bank campaign. As a division chief in the Africa region put it, Narmada had become a 'four letter word' . . . managers in other parts of the Bank reinforced their signals to staff that environment and resettlement should not be ignored or fudged.

The Bank clearly attempted to seize the opportunity to turn around the negative image that the anti-Narmada and multilateral development banks campaigns gave it by unveiling its new environmental management paradigm at the 1992 United Nations Conference on Environment and Development.[48]

The new information disclosure policy and the Inspection Panel (established to assess potential violations of policies with respect to large-scale development programs) instituted by the World Bank in 1993 were also a direct result of the institution's experience with the Narmada Projects and from the ongoing campaigns conducted against the Bank by transnationally allied critics.[49] These reforms had immediate effects. The Inspection Panel's first decision was to mandate the withdrawal of Bank support for Nepal's proposed Arun Hydroelectric big dam project. The Bank also established its Public Information Center to ensure greater availability of Bank documents based on the information disclosure policy the following year. Nongovernmental organizations from the International Narmada Action Committee and the increasingly organized transnational anti-dam network, utilizing such tactics as pressuring their governments to withhold funds until the Bank reformed itself, were once again critical motivators of these changes.

The transnational anti-dam network's concerted focus on the World Bank by the mid-1990s, in particular, is further illustrated by the support for the Manibeli Declaration. As activist Patrick McCully notes:

> this document, drawn up by IRN in coordination with colleagues in India and elsewhere, was submitted to then World Bank President Lewis Preston in September 1994 during the Bank's 50th anniversary celebrations. It calls for a moratorium on World Bank funding of large dams until a number of conditions are met including

the establishment of a fund to provide reparations to people forcibly evicted without adequate compensation, improved practices on information disclosure and project appraisal, and an independent review of the performance of all dams built with World Bank support. The Manibeli Declaration was endorsed by 326 groups and coalitions in 44 countries.

If the member organizations of coalitions are counted separately, the number of individual endorsements for the Manibeli Declaration equaled more than 2,000 from all regions of the world.[50] In that same year, Bruce Rich of the Environmental Defense Fund also published his book, *Mortgaging the Earth: The World Bank, Environmental Impoverishment, and the Crisis of Development,* which leveled a scathing critique on the Bank, especially against its assistance for big dams and other large-scale development projects.

The World Bank responded to the ever-mounting pressure from its transnationally allied critics when its Operations Evaluation Department (OED) began the first ever systematic internal review of big dam projects supported by the Bank. The OED draft report entitled, "The World Bank's Experience With Large Dams: A Preliminary Review of Impacts," was completed on August 15, 1996. The report unequivocally acknowledged the role that external pressure, and particularly the transnational campaign against the Narmada Projects, played in catalyzing reforms at the Bank and motivating the study itself.

> Large dams used to be synonymous with modernization and development. But in the 1970s and 1980s, their adverse indirect and secondary impacts became the target of public criticism. As a result, starting in the World Bank, new policies and standards emerged to help avoid or mitigate the adverse environmental and social consequences of large dams in developing countries.
>
> The advent of these new directives did not still public controversy, partly because of serious implementation problems in a few visible cases and also because, by then, the debate had turned bitter and polarized.
>
> The large dams controversy reached a new peak and focused more directly on the Bank in the 1990s, mainly as a consequence of the Bank-financed Narmada (Sardar Sarovar) projects in India . . . The Bank's role in financing this project and its inadequate management of the resettlement and environmental aspects before India's cancellation of the Bank loan in 1993, disturbed and alienated many environmentalists, and intensified their opposition to large dams. . . .
>
> The controversy surrounding large dams has made potential borrowers reluctant to approach the World Bank and other development agencies for assistance. . . .[51]

The passage clearly corroborates the hypothesis that critics had been increasingly successful in preventing Bank support for big dam projects.

In fact, this report, produced by World Bank staff, also stated that while 26 Bank-supported, dam-related projects were being completed annually between 1970 and 1985, the Bank's involvement in projects that included the construction of a dam declined to about 4 per year from 1986 to 1995.[52] The cumulative effect of transnationally allied anti-dam activity on the World Bank is further demonstrated by the decline in Bank funding for dam projects from the late 1970s on. As shown in Figure 6.1, Bank funding fell from nearly $11 billion in the five-year period between 1978 and 1982 to an estimated $4 billion between 1993 and 1997.

But the OED analysis, based on a desk assessment of fifty World Bank–assisted projects, determined that big dams continued to be effective and thus desirable development initiatives. The report argued that, "mitigation of the adverse social and environmental consequences of large dams would have been both feasible and economically justified in 74 percent of the cases." It thus concluded that, "These results go a long way to help answer the basic question: 'Should the World Bank continue supporting the development of big dams?' Based on the sample reviewed, the answer is a conditional yes, the conditions being that: (i) the projects comply strictly with the new Bank guidelines; and (ii) the design, construction and operation of new projects take into account the lessons of experience."

The OED report acknowledged that limitations in this preliminary analysis were inevitable and proposed convening a representative group of stakeholders to review it and formulate plans for a second phase. At the workshop, members of the transnational anti-dam network not only reiterated their view that the OED report was methodologically flawed and biased, but also renewed the demand for an independent comprehensive review of the World Bank's experience with big dam projects.[53] That meeting unexpectedly generated a proposal for the establishment of an independent world commission to comprehensively review all big dams that had been built worldwide, rather than just those that had involved the Bank. The transnational anti-dam network seized the opportunity, agreed to this proposal, and the process for forming a World Commission on Dams was initiated.

TRANSFORMING THE INTERNATIONAL BIG DAM REGIME

The transnational anti-dam network continued to expand in membership, deepen its activities and become further consolidated during the 1990s as the various reforms at the World Bank were being adopted. The publication in 1996 of IRN campaigns director Patrick McCully's thorough critique of big dam pro-

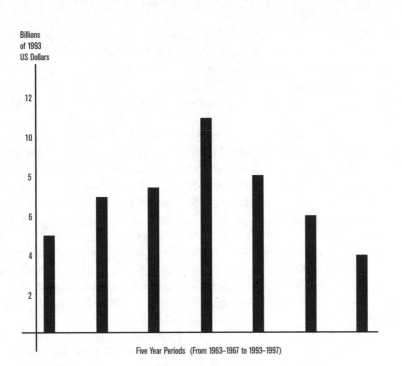

Figure 6.1. World Bank Funding For Big Dams. Figure based on Sklar and McCully (1994), 10. The figure for 1993–97 is a revised estimate based on the latest available sources on the Bank's lending portfolio.

jects around the world, *Silenced Rivers: The Ecology and Politics of Large Dams*, further contributed to the growing understandings and connections among members of the transnational anti-dam network, in the same way that the Goldsmith and Hildyard volumes had done earlier. The book offered an examination of the technical, economic, social, and environmental costs of big dams and their failure to deliver on the benefits expected from them, with a political analysis of the uneven distribution of the costs and benefits among different groups generated by these projects. It also gathered together a brief overview of various big dam conflicts and campaigns from all over the world.

A crucial factor in the strengthening of the transnational anti-dam network at this point was the First International Conference of Peoples Affected by Dams, held in Curitiba, Brazil, March 11–14, 1997. The conference resulted in the strengthening of linkages, the clarification of common goals, and the honing of coordinated strategies within the transnational anti-dam network. It came at an opportune time, as well, exactly one month before the World Bank and the

World Conservation Union (IUCN) convened the meeting of representative stakeholders to review the OED desk report that had analyzed the effects of big dams funded by the Bank. The members of the transnational anti-dam network who attended the latter meeting were extremely well organized, and empowered, to press their demands precisely because many had just participated in the Curitiba conference.

The idea for an international meeting of dam-affected people, and not just dam critics, originated at an annual meeting of MAB, the Brazilian Movement of People Affected by Dams in 1993 (see chapter 5). Members of MAB had been frustrated by the dynamics of the United Nations Conference on Environment and Development and parallel meetings of nongovernmental organizations that had been held the year before in Rio de Janeiro, Brazil. From the perspective of MAB, people who bore the brunt of and were the most marginalized from extant development strategies and projects had not been present at the meetings. MAB thus decided to plan a conference in which the people most directly and negatively affected by big dam projects would be core participants. MAB contacted IRN, which by then had established an office in Brazil, to assist in organizing the event.[54]

The process of organizing The First International Conference of People Affected by Dams was as important in strengthening the links within the transnational anti-dam network as the meeting itself. An international coordinating committee was formed at a preparatory meeting held in Itamonte, Brazil, headed by MAB and consisting of Chile's Biobío Action Group (GABB), the European Rivers Network (ERN), India's Save the Narmada Movement (NBA), and IRN from the United States. The goals established at the meeting were to consolidate the transnational network of people affected by dams and dam-critics, to seek reparations for damages suffered by dam victims, to assure that affected people had an effective voice in decision-making processes, and to educate the public on the true costs and benefits of dams.[55]

The conference itself added further momentum to the consolidation of the transnational anti-dam network. A total of 98 dam-affected people and dam critics speaking more than 12 languages from 17 countries including Argentina, Bolivia, Brazil, Chile, France, Germany, India, Lesotho, Mexico, Norway, Paraguay, Russia, Spain, Sweden, Switzerland, Taiwan, Thailand, and the United States attended the meeting. In a series of plenary and working group sessions, they shared their experiences, failures, and successes.[56] As one attendee later wrote, "Many of the participants experienced an end to the isolation that they feel in their local and regional fights, and as a result, they reported a renewed strength that they could carry back to their communities."[57] Mathato Kthis'ane of the Highlands Church Action Group from Lesotho stated that:

We have been struggling to help the dam-affected people from the Lesotho Highlands Project without knowing what other groups in other countries are doing and how. You can be told that what is happening in your country is happening all over the world, but until you actually spend time with people from other places, you can't really understand this. From this conference, I feel that I should be brave and go back home and talk to the communities, make them feel and know that they are not the only ones, that they should stand on their own feet and try to fight this big problem that they are facing.[58]

The empowering effect of the increased sense of solidarity and various strategies learned from the meeting was echoed by a number of participants.[59]

The conference participants drafted the Declaration of Curitiba, which built on the norms and goals espoused in the San Francisco Declaration of 1988 and the Manibeli Declaration of 1994, and reaffirmed the right to life and livelihood of people affected by dams.[60] The preamble stated that, "We, the people . . . gathered in Curitiba, Brazil, representing organizations of dam-affected people and of opponents of destructive dams, have shared our experiences of the losses we have suffered and the threats we face because of dams. Although our experiences reflect our diverse cultural, social, political, and environmental realities, our struggles are one." It argued that the projected benefits and costs of dams have not matched actual reality and identified a set of powerful actors: international lenders and credit agencies, private companies and corporations, engineering and environmental consultants, technocrats and politicians, as being responsible for the centralized and undemocratic promotion of these projects.

The Curitiba Declaration further consolidated the set of shared norms and goals of the transnational network. The primary principle espoused was the opposition to the construction of any dam that was not approved by the affected people after an informed and participatory decision-making process. Correspondingly, the declaration included the demand for a moratorium on dam building until: (1) there was a halt to all forms of violence and intimidation against people affected by dams and organizations opposing dams; (2) reparations were negotiated with the millions of people whose livelihoods have already suffered because of dams; (3) actions were taken to restore environments damaged by dams—even when this requires removal of the dams; (4) the rights of indigenous, tribal, semi-tribal, and traditional populations were fully respected; (5) an independent commission was established to conduct a comprehensive review of all large dams supported by international agencies, subject to the approval and monitoring of representatives of the transnational anti-dam network, and its policy conclusions implemented; (6) similar independent and comprehensive reviews were established for each national and regional agency which has supported the building of large dams, and their policy conclusions implemented; (7) energy and freshwater policies were implemented which encourage the use of appropriate and sustainable technologies and management practices which guarantee equitable access to basic needs; and (8) democratic and effective public control of electricity and water entities was ensured.

The Curitiba Declaration closed with a pledge by its signers to consolidate the transnational anti-dam network and to intensify the struggle against dams until these demands were met. To symbolize the growing solidarity among anti-dam groups around the world, the document identified March 14, the Brazilian Day of Action Against Dams and for Rivers, as the International Day of Action Against Dams and for Rivers, Water, and Life. A Curitiba Steering Group, consisting initially of the five organizations that took the lead in promoting the First International Conference, was established to coordinate the activities among members of the transnational anti-dam network, and to organize the First International Day of Action Against Dams and for Rivers, Water and Life. Subsequent press articles reported that "events highlighting rivers in crisis and promoting their sustainable and equitable management" occurred in more than 20 countries around the world on March 14, 1998, as part of the First International Day of Action.[61]

As a result of the First International Conference and First International Day of Action, members of the growing and consolidating transnational anti-dam network were extremely focused and organized for the World Bank—World Conservation Union (IUCN) Workshop, "Large Dams: Learning from the Past, Looking to the Future," held in Gland, Switzerland, the following month on April 10 and 11, 1997. In fact, IRN had been contacted by IUCN to propose possible representatives of dam critics and dam-affected groups who should be invited to the Gland workshop. Not surprisingly, the names IRN proposed were various members of the increasingly coordinated transnational anti-dam network, including Save the Narmada Movement in India and the Brazilian Movement of People Affected by Dams.[62]

The Gland workshop was promoted by the World Bank and IUCN as an attempt to bring "together leading experts and representatives of major stakeholder groups . . . to initiate an open and transparent dialogue," on the future of large dams around the world. At the workshop, the 39 participants representing governments, the private sector, international financial institutions, professional associations, scientific communities, nongovernmental organizations, and social movements addressed three issues: critical advances needed in knowledge and practice, methodologies and approaches required to achieve these advances, and proposals for a follow-up process involving all stakeholders.[63]

Private sector participants attended the workshop because they "saw financial, as well as public relations benefits to winning environmentalists over to their side. 'With a multimillion dollar dam, the costs of long political delays are enormous,' said Mr. Jan Strombland of ABB, the engineering group which had received a beating from pressure groups for its involvement in the Bakun Dam." Most government representatives were interested in finding means to achieve the water and power benefits of big dams while avoiding the social and environmental costs associated with building them.[64] Leaders from international professional and industry associations like the International Commission on Large Dams and International Hydropower Association were perhaps the most vocal in their defense of big dam building.

But members of the transnational anti-dam network, based on the demands articulated in the Curitiba Declaration, coordinated their activities at the Gland workshop to advocate for an independent and comprehensive review of big dams supported by international funding agencies. IRN Campaigns Director Patrick McCully was asked to present a review of the OED report on World Bank dam building at the workshop. Not surprisingly, he leveled a thorough critique:

> The OED review does not assess the actual performance of the projects it covers, is based on flawed methodology and inadequate data, and displays systematic bias in favor of large dam building. Its conclusions must be rejected as untenable. The extremely poor quality of the OED review strongly indicates that OED is not a suitable body to entrust with the task of undertaking a comprehensive, unbiased and competent review of World Bank lending for large dams . . . For a review of World Bank-funded dams to be seen as credible and unbiased it must be done by a commission of eminent persons independent of the World Bank. The commission must be able to command respect and confidence from all parties involved in the large dams debate and must seek and utilize substantive input from critics of current dam-planning and building practices.[65]

As an article in the Financial Times later acknowledged that, "Many participants agreed with the claims of Mr. Patrick McCully," that the report, "was based on insufficient information about experience on the ground."[66]

The forcefulness of the critique and the coordinated political pressure from representatives of the transnational anti-dam network resulted in agreement to establish an independent World Commission on Dams (WCD). World Bank representatives proposed that the WCD review all big dams built around the world, and not just those projects with Bank involvement, an idea that was accepted by all participants. Private sector participants and other big dam proponents urged that the WCD also generate guidelines for the future in addition to reviewing past experience.

Correspondingly, the mandate of the WCD stated that, "The overarching goals of the Commission are to: (a) review the development effectiveness of dams and assess alternatives to them, and (b) formulate internationally accepted standards, guidelines and criteria for decision-making in the planning, design, construction, monitoring, operation and decommissioning of dams related to the sustainable development and management of water and energy resources." In order to ensure the independence of the WCD, it was agreed that the World Bank and IUCN would maintain their roles as initiators but would not interfere with the work program of the Commission and that funding would be sought out from a wide array of sources.[67]

The subsequent conflicts over the selection of commissioners further attested to the considerable power that the transnational anti-dam network wielded. In particular, the big dam critics on the reference group criticized the interim secretariat of the WCD for allowing pro-dam interests to minimize representation of nongovernmental organizations and affected peoples' groups in the initial list of commissioners produced in November 1997, based on the negotiations among stakeholders. The threat that these groups would withdraw from the WCD even before it even started compelled the interim secretariat and the consensus choice for chairman, Professor Kadel Asmal, who was the South African Minister of Water Affairs, to convene an emergency meeting of a small number of representatives of the various stakeholder groups in early 1998.

The result of this meeting of various representatives that ostensibly represented the fully array of stakeholders in the conflicts over big dams was the selection of a vice-chair proposed by the critics, replacement of one pro-dam commissioner unacceptable to the anti-dam groups with another pro-dam commissioner who was considered less problematic, and the addition of another commissioner, Medha Patkar, the activist leader of India's Save the Narmada Movement. The final composition of the WCD thus ostensibly included four representatives of nongovernmental organizations and social movements who were big dam critics, four commissioners who seemed to be moderates on the issue, and four who were considered to be proponents.[68]

Both the decline in World Bank funding and reform in Bank policies as well as the creation, mandate, and composition of the WCD provide compelling evidence of major shifts in the norms, principles, procedures, and structures of the international big dam regime. When this regime was initially started with the formation of ICOLD in 1929 and consolidated with the first "development decade" of the 1950s, representatives of transnationally allied nongovernmental organizations, grassroots groups, and social movements were not (active) participants. But these actors were critical players in increasingly altering the dynamics of big dam building and development from the 1970s onward and in constituting the governance structure and shaping the goals of the WCD. More-over, globally spreading social and environmental norms had been enshrined as being of equal value with historically hegemonic economic priorities in formu-lating the meanings, goals, and evaluative standards related to big dam building and development that were included in the agreed-upon mandate of the Com-mission.[69]

COMPETING EXPLANATIONS FOR THE CHANGES IN BIG DAM BUILDING

In this study I endeavored to examine the dramatic rise and fall of big dam building around the world during the 20th century, and to specifically explain the especially puzzling decline since the 1970s. The analysis offered in chapter 1 revealed that the former trend was attributable to the presence of powerful transnationally allied big dam proponents and the constitution of an interna-tional big dam regime supporting and legitimating these projects. The latter trend was substantially accounted for by the unexpectedly sizeable reduction in the construction of these projects across the third world.

The unexpected decline in developing countries was conditioned, I have ar-gued, by the emergence and growing strength of transnationally allied opposi-tion to big dams led by nongovernmental organizations, social movements, and grassroots groups interacting with the global spread and multi-level adoption of norms on the environment, human rights, and indigenous people. This com-bination of transnational advocacy and norm institutionalization contributed to the significant reduction and/or the substantial reform in big dam building within countries across the third world when domestically buttressed by orga-nized and sustained opposition in the context of a democratic regime.

This full set of interacting factors was most clearly visible in India from the 1970s to the mid-1990s as demonstrated in chapters 2, 3, and 4. The growing au-thoritarianism of India's polity, however, increasingly minimized the impact that could occur by the end of the 1990s. Big dam building in China, examined in chapter 5, provided a vivid and contrasting case in which the relative absence of both the two critical domestic-level factors of democracy and social mobi-lization posed a powerful barrier to the presence and impact of transnational political action and globalizing norms.

The analysis in chapter five on variation and changes with respect to big dam projects in Brazil is especially important. In addition to offering further evi-dence in support of the core argument of this book, it revealed that affected peo-ples, domestic nongovernmental organizations, and allied foreign critics often join and contribute to pro-democracy movements partly because these institutional contexts offered greater political opportunities to alter extant big dam building practices and partly because democratic norms of transparency, participation and accountability were aspects of the broader vision of development that motivated these actors in the first place.

Indeed, it was discovered that the opposition to big dam building was a means for the transnationally allied promotion of domestic democratization in many contexts, because demands for open access to information, adherence to civil rights and political liberties, and greater participation in decision-making processes were central to most anti-dam campaigns. This was certainly true in the case of Indonesia, as was shown in chapter 5, for example when a wide array of civil society actors focused on the Kedung Ombo Project to protest against not only the abuses, but also the very existence of the domestic authoritarian regime. The persistence of Indonesia's authoritarian regime until the late 1990s meant that the combination of factors required for generating more dramatic shifts in big dam building was not operative in that country.

Yet the installation of more democratic institutions does not by itself ensure the halting or substantial reform of big dam building either, as illuminated in chapter 5 by the case of South Africa more generally and the Lesotho Highlands Water Project(LHWP) more specifically during the 1990s. The lack of broad-based mobilization by affected peoples and supporting domestic nongovernmental organizations, despite the transitions to democracy in Lesotho and South Africa, resulted in completion of the first and continued work on the second of the five big dams that constitute the LHWP as well as relatively unabated construction of big dams in South Africa.[70]

The building of effective transnational coalitions, not to mention durable transnational networks, across wide geographical, cultural, and linguistic boundaries is quite difficult. For these forms of transnational advocacy to be buttressed by the institutionalization of supportive global norms in countries with persistent democratic institutions and linked to strong domestic social mobilization raises the threshold for impact even higher.

But the progressive consolidation of the transnational anti-dam network (and related networks on indigenous peoples, human rights, the environment, anticorruption, etc.) and the increasing transnational solidarity among domestic dam-affected groups examined in chapter 6 may have structurally altered the landscape for big dam building in the 21st century. As L. David Brown and Jonathan A. Fox suggest, "In the long run, such (interactions) may generate the social capital—reflected in the proliferation of bridging organizations and individuals capable of building relationships and trust among diverse actors—required to construct effective transnational movements."[71]

Hence, if transnational movements are sets of actors linked across country boundaries that have the capacity to generate grassroots social mobilization in multiple locations to publicly and nonviolently influence social change, then the First International Conference of Peoples Affected by Dams, and the subsequent International Days of Action Against Dams and for Rivers, Water and Life might signal the formation of a transnational movement in this issue area.[72] While some analysts are skeptical of the notion of a transnational movement, the deepening of transnational links between grassroots peoples affected by big dams and the emergence of joint social mobilization may empower and thus increase the effectiveness of previously weak and disorganized domestic groups.[73]

The continuing global spread and institutionalization of supportive norms at various levels of authority from local implementing agencies to international organizations—and emergence of novel forms of global governance like the WCD— may also increase the likely effect of these transnational structuration mechanisms. Indeed, it is quite clear that there has been a rise and global spread of norms on participatory and democratic governance that are both shaping and being shaped by various transnational networks.[74] Correspondingly, domestic level conditions may become more conducive if democratic regimes are consolidated in the scores of countries that underwent transitions from authoritarian rule from the 1970s on and other polities (such as those in Indonesia and Nigeria) also potentially become more democratic.

Competing explanations for the decline in big dam building were proposed in chapter 1 and examined in the subsequent chapters of the book. Perhaps the major alternative account is that the falling rate of big dam building was caused by the decline in sites available for the construction of these projects. The cases analyzed in this study, namely India, Brazil, Indonesia, China, and South Africa were extremely tough tests for the general argument posed in this study relative to this alternative explanation, because these five have been among the top big dam building countries in the world during the second half of the 20th century. Indeed, there is no question that the declining availability of sites has contributed to the decline in the numbers of big dams built in these and other countries.

But as demonstrated in chapters 2 through 5, even in these countries, numerous sites for constructing big dams remain available. In addition, perhaps with the exception of China, the number of proposed big dam projects has not declined as much or as quickly (if at all) as the number of completed big dam projects, suggesting that other factors besides site reduction have been operative. Other third world countries where big dam building was much less expansive, and thus where the percentage of sites utilized compared with the total number of sites available is still quite high, have demonstrated similar patterns in the construction and/or reform of these projects. Indeed, in the case of Indonesia, where the lack of a democratic regime undermined the effects of transnational contestation and globalizing norms, the number of projects completed actually increased from the 1970s through the 1990s.

Another major competing explanation that linked changing economic and financial conditions to the decline in big dam building can also not be completely discarded based on the evidence presented in this study. The relative economic viability of conventional alternatives, such as some types of thermal power plants due to (the temporary) lower prices for natural gas during the 1990s, for example, certainly reduced some of the interest in big dam projects from time to time. In addition, such factors as economic recession in numerous donor countries, donor fatigue more generally, debt and economic crises within developing and transition economies, among others—all have individually and interactively combined to reduce the overall financing available for big dam building.

Yet, the analysis conducted in this study suggests that public funding agencies (such as multilateral development banks, bilateral donors, and export credit agencies) as well as third world state authorities continue to want to finance the building of these projects. Indeed these actors would have done so much more had it not been for the rising costs from intensified transnationally allied opposition and associated globalizing environmental, human rights, indigenous peoples, and other norms. Moreover, with the declining availability of public funds for these projects, proponents have increasingly sought domestic and international private financing to fill the gap.

Indeed, many states and donors, including the World Bank and specifically its private sector arm, the International Finance Corporation (IFC), increasingly began to provide risk guarantees to potential private investors in big dam projects by the late 1990s.[75] As a proponent of hydropower development argued at the time, "The issue for project owners, constructors, and suppliers no longer is: should we use private money for hydro development? The question is: how do we mobilize the private sector sufficiently to develop all the new hydroelectric capacity we need in the world?"[76]

Not only did the privatization and liberalization of big dam building potentially increase the financing available for big dam projects; these trends also began to undermine the improved social, environmental, and decision-making policies and procedures that were generated since the 1970s. For example, the IFC, MIGA, export credit agencies, and private sector firms were arguably less transparent, participatory, and accountable or supportive of human rights, indigenous peoples, and environmental norms by the end of the 1990s than multilateral development agencies such as the World Bank. Not surprisingly, the transnational anti-dam network correspondingly began to focus more of its attention directly on these other transnationally allied big dam proponents. The privatization and liberalization of big dam building was also a key issue on the agenda of the World Commission on Dams.

This study has not devoted considerable analytic space to examining the counter-activities of various pro-dam interests, such as the more recent promotion of private financing, ownership and/or management of big dams.[77] Pro-dam interests were not actively organized and they were not as politicized compared with similar types of actors in other issue areas, such as trade or oil. This is partly because big dam building had become so legitimated, naturalized, and habituated as appropriate behavior by the 1950s and 1960s that organizing politically was unnecessary. However, the increasing loss of big dam building's unquestioned acceptance as appropriate practice around the world from the 1970s on motivated pro-dam actors to become more organized to defend their interests as they understood them.[78]

Professional technical associations such as the International Commission on Large Dams (ICOLD), International Commission on Irrigation and Drainage (ICID), and the International Hydropower Association (IHA), not to mention numerous multinational corporations, international donor agencies, and various domestic interest groups, began to mount stronger transnationally allied campaigns of their own to promote these projects during the 1990s.[79] In fact, the IHA was purposefully created in the mid-1990s to advocate the cause of hydropower as a clean and renewable energy that ostensibly contributed to reductions in global climate change. But this environmental-sustainability-based justification for promoting hydropower was very different from the big-dams-are-development argument that had historically become established and that most pro-dam actors continued to advance through the turn of the 21st century.

Certainly, the active involvement of these big dam proponents in the World Commission on Dams was indicative of their growing efforts to balance big dam critics. In addition to their claims that big dams had been and continued to be essential for economic development, pro-dam groups also integrated arguments that these projects were or could easily become instruments for the realization of social and environmental goals in broader notions of sustainable development.[80] And these proponents increasingly began to charge big dam critics for being undemocratic themselves.

In fact, the representativeness and accountability of transnational anti-dam organizations, coalitions, and networks remained problematic throughout this period of dramatic change. While much of the moral authority of transnationally allied contestation was based on representing dam-affected peoples, positions and tactics were often articulated by nongovernmental organizations and individuals that were not linked directly or clearly accountable to dam-affected peoples. More resource-endowed nongovernmental organizations tended to be much more influential in shaping campaign strategies and gaining access to decision makers. This asymmetry was exacerbated because procedures for ensuring that the perspectives of all allied groups were given equal weight and mechanisms for resolving differences between different groups remained underdeveloped.[81]

Furthermore, because critical shortages and inequities of drinking water, food (that irrigated agriculture might produce), and electricity remained even after big dams were halted, transnational campaigns contesting these projects may have unintentionally contributed to the adoption of greater numbers of other conventional alternatives. The transnationally allied opponents of big dams thus may not have been as effective in achieving their broader vision of sustainable development, if the negative social and environmental effects from these conventional alternatives, such as the global warming effects of coal-fired thermal plants, are comprehensively analyzed.[82]

In fact, while transnationally allied opposition was extremely effective in substantially reforming and halting big dam building, it had much less impact on the promotion of alternatives for sustainable development. This is not surprising given that the relative effort spent on opposing big dam projects during the 1970s to the 1990s was much greater, although these critics argued more and more for such options as demand-side management, traditional water-harvesting practices, solar and wind power projects and others over time. The range of potential options for the sustainable development and management of water and energy resources, and the transnational political economy and institutional arrangements that might facilitate their adoption and implementation, thus remained critical dilemmas for the 21st century.

DAMS, DEMOCRACY, AND DEVELOPMENT IN TRANSNATIONAL PERSPECTIVE

The dramatic and inter-linked changes in big dam building and the international big dam regime indicate that a broader transformation in the transnational political economy of development has occurred. This is because big dams are activities and symbols that reflect, are conditioned by, and shape larger dynamics of development, dynamics that are transnational in nature. Correspondingly the more general implications of this book contribute to theoretical and empirical advances as well as future research agendas in the scholarly fields of development and transnationalism.

This study offers important insights into the long-standing social scientific debates on the relationships between institutions and development. The evidence presented in this book supports the argument that democratic institutions increase the likelihood that transnational advocacy, certain types of globalizing norms, and domestic mobilization reduce and/or substantially reform big dam building. Many would argue that such changes around these projects contributed to sustainable development, if sustainable development is understood as longer-term progress toward greater public participation, political accountability, social equity, and environmental sustainability. Thus, it could be concluded that democratic institutions are more conducive to the promotion of (sustainable) development.

However, if more and not fewer big dam projects were considered necessary for development, then an authoritarian regime might be posited as more conducive. Suppose economic growth through increased agricultural production and industrialization is the prevailing vision of development, and electricity, irrigation, and other outputs of big dams are considered necessary to achieve that (economic) development. Then authoritarian regimes would increase the likelihood of development by minimizing the effects of transnational contestation, globalizing norms, and domestic mobilization that might have otherwise prevented these projects from being built.

This suggests that scholarly research investigating the effects on development of democratic versus authoritarian regimes—or a wider range of institutions including the rule of law or enforceable property rights or certain types of norms, or other factors such as "good governance," "social capital," levels of instability, savings, or investment more broadly—that operationalizes development in different ways would contribute to advances in knowledge. Scholars can no longer, if they ever could, assume that the conceptualization of development they adopt will be universally acceptable.[83]

In other words, a general theoretical implication of this study is that the causes of development depend on the meanings of development, at least in part. In other words, development is socially produced and reproduced as well as socially constructed and reconstructed. Moreover, in order to (causally) reform or transform development, the meanings of development may have to be contested and reconstituted.

In fact, opponents had to challenge the taken-for-granted equivalence between big dams and development partly by critiquing the vision of development that legitimated these projects (and offering alternative notions). Thus, explicating, interpreting, and understanding (the genealogies, structures, and effects of different socially constructed) meanings of development are in and of themselves important scholarly tasks.[84]

Constructivist scholarship based on such understandings offers a productive and dialogical option in relation to important but ultimately unsatisfactory positivist and interpretivist approaches to development studies. Indeed Frederick Cooper and Randall Packard suggest in their incisive sociological interrogation of knowledge on development in the social sciences that two more extreme versions of these alternative approaches increasingly polarized the study of development from the 1980s on. They write: "One set might be called ultramodernist. It consists of economic theorists who insist that the laws of economics have been proven valid," and scholarship that considers the basic assumptions and tools of economics as of predominant utility in research on development. In contrast, a "second set is postmodernist. This group sees development discourse as nothing more than an apparatus of control and surveillance . . . a 'knowledge-power regime', that scholars should critically excavate and unmask."[85] This book offers an example of how the insights of these contending scholarly approaches can be drawn on and complement other epistemological, ontological, theoretical, and methodological perspectives to illuminate and explain development dynamics in a constructive constructivist way.

Furthermore, development is not just an essentially contested concept for scholarly research in isolation; development is also an intensely conflict-ridden empirical phenomenon.[86] Big dams as development activities and symbols provide ample proof of this claim. Indeed, this book offers compelling evidence that suggests that struggles over development occur not just from inequalities in control or outcomes but also from contending understandings about the causes and meanings, goals and evaluations of development.[87]

Moreover, conflicts over development and development dynamics more generally are (perhaps increasingly but certainly always have been) transnationally constructed. The pervasiveness of transnational interactions among 'agents' and 'structures', causes and meanings, can be seen in any issue area that is linked to or involves development, from natural resources and mining to trade, debt, finance and investment, to humanitarian intervention and peace building. For example, complex sets of transnational structuration mechanisms contributed to the global rise and spread of both state-led import-substituting industrialization development models in the 1950s and '60s as well as the progressive shift towards market-led, neo-liberal strategies from the 1980s on.

This book is therefore also part of a reemergent social scientific research agenda on transnationalism in addition to a contribution to scholarship on development. It contributes to this "new transnationalism" literature by elaborating different types of novel transnational actors and institutions, and by identifying key conditions and mechanisms through which these agents and structures (re-)shape dynamics of development across multiple contexts and levels.

The analysis herein further reveals how the identities of various actors involved in these novel forms of transnational action are themselves altered over time. Big dam critics from the West were primarily environmental conservationists, whereas those from developing countries were primarily focused on social justice issues. Through their interaction over time, these actors not only began to adopt the norms and discourses of their counterparts but novel reframings occurred such as an increasing explication of a "rights and sustainable livelihoods" perspective that was linked to struggles over big dams and debates over development more broadly.[88]

The novel form of transnational contestation around development as reflected by the struggles over big dams analyzed in this study is by no means unique; rather, these struggles have become pervasive aspects of world affairs. As discussed in several places in this book and examined elsewhere, similar sets of transnationally allied actors have been organizing and mobilizing advocacy in a variety of areas including nuclear weapons, women's rights, democratization, land mines, anticorruption, debt-relief, and others.[89]

Indeed, a range of these transnational actors has evolved from informal and short-term coalitions to more formal and institutionalized issue networks and organizations. As a result, they increasingly constitute structural parts of the broader transnational organizational sectors or social fields of security, governance, etc., that parallel and overlap with the transnational field of development. Investigation of the structuring of these transnational organizational sectors and social fields, their empirical implications for world affairs and theoretical implications for social science knowledge is a critical area for further research.

Compared with other scholarship like it, the interactions among such novel "transnationalisms from below" with extant and evolving "transnationalisms from above" with respect to big dam building were also explicated in this book.[90] The latter included the activities of various types of transnational actors such as multinational corporations, international professional associations, and multilateral agencies, among others. These agents interacted to generate and perpetuate the norms, principles, and procedures of what was called an international big dam regime. But these transnationalisms from above, imbricated with the international big dam regime, had also become relatively durable and structural parts of the transnational field of big dam building and thus development, even earlier than the novel phenomena and dynamics that contributed to the transformation of this field.

In sum, this study contributes to the building of a research program on "constructivist transnationalism" for the study of world affairs. The recent wave of scholarship on transnationalism in political science and sociology remains primarily liberal and actor-centric in orientation.[91] This study generates a strong case for a more explicit focus on different types of power (albeit with a much broader notion of power than the neo-realist tradition) as well as structures and structuration processes, in addition to the motivations and strategies of actors in order to explain and understand transnational phenomena and dynamics.

The book correspondingly links emerging propositions about transnational actors with conventional arguments about international regimes and more recent "constructivist" theorizing in international relations. The leading international relations work in the constructivist tradition has predominantly focused on states and the inter-state system, and has not by and large explicitly examined transnational non-state agents, structures, or dynamics.[92] Moreover, as Peter Katzenstein, Robert Keohane, and Stephen Krasner state: "There is a growing body of work in international relations and in security studies but, significantly, not yet in IPE (international political economy) that is self-conscious in conducting empirical research from a constructivist perspective."[93] This book is thus a conscious attempt to contribute an empirically grounded and conceptually rigorous study in the constructivist tradition linked to scholarship on transnational actors in the international political economy issue area.

It does so, however, by drawing and adopting insights from the world society, neoinstitutionalist and other traditions in sociology that have so forcefully explicated and documented the institutionalization of social and organizational fields. But the book also explicates processes and specific mechanisms of norm and field emergence, spread, adoption, and internalization that are generally not identified in this sociological research agenda. Moreover, world society and world system approaches in sociology largely investigate different types of global social structures, whether ideational and cognitive as in the case of the former or materialist and economic in the latter, identify states as the primary units embedded in and conditioned by these structures, and minimize the role of agency whether of state or non-state and transnational actors.[94]

The constructivist transnationalism scholarship promoted in this book offers a potentially productive option to these more institutionalized research programs in sociology and political science for the study of world affairs. This emergent paradigm is based on at least the following three conceptual moves: (1) from international-statist and world-statist to transnationalist (or multiple actors and structures that potentially cross borders and boundaries) theory; (2) from rationalist and structuralist to structurationist ontology; and (3) from positivist and interpretivist to constructivist epistemology. A constructivist transnationalism is exciting because it offers a research program that can inform the study of world affairs given the permeability of borders and boundaries and import of cross-border and cross-boundary factors to dynamics from the local to the global levels.[95]

Several avenues for future research in a constructivist transnationalism scholarly agenda would be well worth pursuing. The first is continued comparative investigation across issue areas and institutional contexts of transnational structuration "from below" to further improve understanding of the forms it takes and the conditions under which it can generate change. Correspondingly, comparative examination across issue areas and institutional contexts of the responses of transnationally allied dominant groups such as multinational corporations and business associations and/or various types of governments, among others, to these challenges from below would be of great utility in illuminating critical interactions in world affairs.

Research on the shifts occurring in extant international organizations and emergent transnational governance forms would also be extremely valuable.[96] Perhaps of greatest interest and importance for a research program on "constructivist transnationalism" would be to systematically examine and analyze over time and across space various transnational phenomena (agents and structures) and structuration dynamics (meanings and causes)—transnational terrorist and criminal networks, transnational religions and communities, transnational financial flows, transnational migration, transnational nongovernmental organizations and social movements, transgovernmental relations, transnational cultural models, etc.—and how these fields (re)shape identities, processes, and outcomes from the local to the global levels around the world. The potential significance of such a constructivist transnationalism research agenda is great given the knowledge it could generate for the complex and dynamic world of the 21st century. This book, with all its remaining flaws, will hopefully inspire more and better scholarship of this kind.

❖

NOTES

1. TRANSNATIONAL STRUGGLES FOR WATER AND POWER

1. International Commission on Large Dams, "Don't deny us our dams," *International Water Power and Dam Construction* (December 1994): 14–15.
2. Edward Goldsmith and Nicholas Hildyard, eds., *The Social and Environmental Effects of Large Dams,* vol. 1 (Wadebridge: Wadebridge Ecological Centre, 1984), 345.
3. Interview, New Delhi, 27 April 1996 (translated from Gujarati).
4. Interview, Baroda, Gujarat, 1 February 1996 (translated from Hindi).
5. See Mahesh Pathak, ed., *Sardar Sarovar Project: A Promise for Plenty* (New Delhi: Oxford, 1991).
6. The English language term in India for "indigenous peoples" is "tribal peoples." Tribals, or adivasis, constitute approximately one-tenth of the total Indian population or about one hundred million people.
7. See Claude Alvares and Ramesh Billorey, *Damming the Narmada: India's Greatest Planned Environmental Disaster* (Penang: Third World Network, 1988).
8. See R. Sarkaria, ed., *Report of the Commission on Centre-state Relations* (New Delhi: Government of India, 1987), 489–490.
9. I include both "large" and the even more massive "major" dams in the category of "big dams." According to instructions of the International Commission on Large Dams, domestic dam agencies can report dams with heights over 15 meters, and those of 10–15 meters if they meet other technical requirements, as "large dams." "Major dams" meet one or more of the following requirements: height of over 150 meters, volume greater than 15 million cubic meters, reservoir storage of more than 25 cubic kilometers, and/or electricity generation of more than 1,000 megawatts. See ICOLD, *World Register of Large Dams* (Paris: ICOLD, 1988), and T. W. Mermel, "The World's Major Dams and Hydroplants," in *International Water Power and Dam Construction Handbook* (Sutton, U.K.: IWPDC, 1995).
10. See Lori Udall, "The International Narmada Campaign: A Case of Sustained Advocacy," in *Toward Sustainable Development: Struggling Over India's Narmada River,* ed. William F. Fisher (New York: M. E. Sharpe, 1995).

11. "World Bank Staff Appraisal Report, Supplementary Data Volume," pt. 1 (Washington, D.C.: World Bank, 1993)

12. Bradford Morse and Thomas Berger, *The Sardar Sarovar Projects: Report of the World Bank Independent Review* (Canada: Probe International, 1993).

13. They have also been empowered and contributed to globally spreading norms on good governance such as the control of corruption, transparency, participation, and accountability.

14. A World Bank consultant wrote in 1987 that, "most scenarios of future developments in water resources agree that ultimately, say by the mid-21st century, all of the runoff in all of the world's rivers must be stored by reservoirs (from the construction of big dams) or other methods." See K. Mahmood, *Reservoir Sedimentation: Impact, Extent and Mitigation* (Washington, D.C.: World Bank; 1987), 6.

15. See note 9 for definitions of big, large, and major dams.

16. Norman Smith, *A History of Dams* (London: Peter Davies, 1971).

17. See Nicholas Schnitter, *A History of Dams: The Useful Pyramids* (Rotterdam: Balkania, 1994), 158.

18. See John Waterbury, *Hydropolitics of the Nile Valley* (Syracuse, N.Y.: Syracuse University Press, 1979).

19. See A. D. Rassweiler, *The Generation of Power: The History of Dneprostroi* (Oxford: Oxford University Press, 1988).

20. See Donald Worster, "The Hoover Dam: A Study in Domination," in *The Social and Environmental Effects of Large Dams,* ed. Edward Goldsmith and Nicholas Hildyard, vol. 2 (Wadebridge: Wadebridge Ecological Centre, 1986), 21.

21. See Marc Reisner, *Cadillac Desert: The American West and Its Disappearing Water* (London: Seeker and Warburg, 1986), and W. Chandler, *The Myth of the TVA: Conservation and Development in the Tennessee Valley, 1933–1983* (Cambridge: Ballinger, 1984).

22. See Patrick McCully, *Silenced Rivers: The Ecology and Politics of Large Dams* (London: Zed Books, 1996).

23. A similar type of international professional, technical, voluntary, and nongovernmental association called the International Commission on Irrigation and Drainage (ICID) was established in 1950 with offices in India. ICID was initially founded by individuals from eleven countries. The number of "member" countries rose to nearly one hundred, fifty years later. ICID's mission has predominantly focused on the transfer of water management technology for irrigated agriculture, making it a very active big dam proponent in the world.

24. Quoted in Worster, in Goldsmith and Hildyard, 21.

25. See Edward Mason and Robert Asher, *The World Bank Since Bretton Woods* (Washington, D.C.: Brookings Institution, 1973).

26. Albert O. Hirschman, *Development Projects Observed* (Washington, D.C.: Brookings Institution, 1967).

27. Leonard Sklar and Patrick McCully, "Damming the Rivers: The World Bank's Lending for Large Dams," in *International Rivers Network Working Paper Number 5* (Berkeley: International Rivers Network, November 1994).

28. See McCully.

29. See Ismail Serageldin, "Toward Sustainable Management of Water Resources," in *Directions in Development Working Papers* (Washington, D.C.: World Bank, 1996), 1–5

30. See Thomas F. Homer-Dixon, *Environmental Scarcity and Global Security* (New York:

Foreign Policy Association, 1993); Miriam R. Lowi, *Water and Power: The Politics of a Scarce Resource in the Jordan River Basin* (Cambridge: Cambridge University Press, 1995); and Ben Crow et al., *Sharing the Ganges: The Politics and Technology of River Development* (New Delhi: Sage Publications, 1995).

31. See Sandra Postel, "Forging a Sustainable Water Strategy," in *State of the World 1996,* ed. Lester R. Brown et al. (New York: W. W. Norton, 1996), 40–59. Several international initiatives were organized in this area during the 1990s, including the Global Water Partnership, World Water Commission, and World Commission on Dams.

32. International Energy Agency, *Energy in Developing Countries* (Paris: IEA, 1995), 3.

33. See World Bank, *Energy Strategies for Rural and Poor People in the Developing World* (Washington, D.C.: Industry and Energy Department, World Bank, 1995).

34. The numbers of big dams being completed decreased partially as a result of the larger and larger size of projects that were initiated over time. For example, 10 major dams were built before 1950, 35 during the 1950s, 64 during the 1960s, and 93 during the 1970s. Figures calculated from Mermel (1993–1996) and ICOLD, *World Register of Dams* (Paris: ICOLD, 1998).

35. ICOLD (1998).

36. See Jose Roberto Moreira and Alan Douglas Poole, "Hydropower and Its Constraints," in *Renewable Energy: Sources for Fuels and Electricity,* ed. Thomas B. Johansson et al. (Washington, D.C.: Earth Island Press, 1993).

37. See ICOLD, Circular Letter 1427, December 1995; Circular Letter Number 1388, 16 January 1995; and Circular Letter 1342, 4 October 1993, Paris: ICOLD, as well as ICOLD (1998). "Under construction" and "starts" can mean anything from receiving formal approval from authorities to actually being in the process of being built with a greater likelihood of completion.

38. This does not include nonconventional and renewable sources of energy generation, such as wind or solar power.

39. See World Bank, *World Development Report: The Challenge of Development* (Oxford: Oxford University Press, 1991) and World Bank, *Meeting the Infrastructure Challenge in Latin America and the Caribbean* (Washington, D.C.: World Bank, 1995a).

40. See World Bank, *World Development Report: Infrastructure* (Oxford: Oxford University Press, 1994).

41. See R. Repretto, *Skimming the Water: Rent-Seeking and the Performance of Public Irrigation Systems* (Washington, D.C.: World Resources Institute, 1986), and E. W. Morrow and P. F. Shagraw, *Understanding the Costs and Schedules of World Bank Supported Hydroelectric Projects* (Washington, D.C.: World Bank Industry and Energy Department, 1990).

42. See Ian Pope, "Solving the Project Funding Puzzle," *Hydro Review Worldwide* (April 1996): 30–32; Robert Goodland, "The Big Dam Controversy: Killing Hydro Promotes Coal and Nukes: Is that Better for Environmental Sustainability?" (paper presented at GTE Technology and Ethics Series, Michigan Technological University, May 1995); and Robert Goodland, "Large Dams: Learning From the Past, Looking to the Future" (paper presented to the IUCN—The World Conservation Union & The World Bank Group Joint Workshop, Gland, Switzerland, 10–11 April 1997).

43. The World Bank estimated that a one-year delay in completion will reduce the benefit-cost ratio by one-third and a two-year delay by over one-half and that costs associated with resettlement can increase project costs up to 30 percent. World Bank, *The*

Bankwide Review of Projects Involving Involuntary Resettlement (Washington, D.C.: World Bank, 8 April 1996).

44. For the United States, see Tim Palmer, *Endangered Rivers and the Conservation Movement* (Berkeley: University of California Press, 1986), and R. Gottlieb, *Forcing the Spring: The Transformation of the American Environmental Movement* (Washington, D.C.: Earth Island Press, 1993).

45. See Lester Brown et al., *Vital Signs, 1995* (London: Worldwatch Institute/Earthscan, 1995); and Anthony Churchill, "Meeting Hydro's Financing, Development Challenges," *Hydro Review Worldwide* (Fall 1994). See also World Commission on Dams, *Dams and Development: A Framework for Decision-Making* (London: Earthscan, 2000).

46. See Sanjeev Khagram et al., eds., *Restructuring World Politics: Transnational Social Movements, Networks and Norms* (Minneapolis: University of Minnesota Press, 2002); Ann Florini, ed., *The Third Force: The Rise of Transnational Civil Society* (Washington, D.C.: Carnegie Endowment for International Peace, 2000); John Boli and George M. Thomas, eds., *Constructing World Culture: International Nongovernmental Organizations since 1875* (Stanford: Stanford University Press, 1999); Jonathan A. Fox and L. David Brown, eds., *The Struggle for Accountability: The World Bank, NGOs and Grassroots Movements* (Cambridge: MIT Press, 1998); Jackie Smith et al., eds., *Transnational Social Movements and Global Politics: Solidarity Beyond the State* (Syracuse, N.Y.: Syracuse University Press, 1997); Thomas Risse-Kappen, ed., *Bringing Transnational Relations Back in: Non-State Actors, Domestic Structures, and International Institutions* (Cambridge: Cambridge University Press, 1995); Matthew Evangelista, "The Paradox of State Strength: Transnational Relations, Domestic Structures, and Security Policy in Russia and the Soviet Union," *International Organization* 49 (Winter 1995): 1–38; David Zweig, "'Developmental Communities' in China's Coast: The Impact of Trade, Investment, and Transnational Alliances," *Comparative Politics* 27, no. 3 (April 1995): 253–274; Patricia Chilton, "Mechanics of Change: Social Movements, Transnational Coalitions, and Transformation Processes in Eastern Europe," *Democratization* 1, no. 1 (Spring 1994): 151–181; Kathryn Sikkink, "Human Rights, Principled Issue-Networks and Sovereignty in Latin America," *International Organization* 47, no. 3 (Summer 1993); Allison Brysk, "From Above and Below: Social Movements, the International System, and Human Rights in Argentina," *Comparative Political Studies* 26, no. 3 (October 1993): 259–285; and Thomas Princen and Matthias Finger, *Environmental NGOs in World Politics: Linking the Local and the Global* (London: Routledge, 1994).

A previous literature on transnationalism dates back to the 1970s. This research program generated a great deal of work and debate but was overshadowed by state-centric and world system approaches that subsequently became dominant in political science and sociology. For political science, see Robert O. Keohane and Joseph S. Nye Jr., eds., *Transnational Relations and World Politics* (Cambridge: Harvard University Press, 1971). The more sociological world systems and "dependency" research program that emerged in the 1960s and 1970s, and focused on the trends, patterns, and affects of transnational capitalist expansion contributed greatly along these lines as well. See generative work such as Fernando E. Cardoso, "Dependency and Development in Latin America," *New Left Review,* no. 74, 83–95; Andre Gundre Frank, *Capitalism and Underdevelopment in Latin America* (Cambridge: Cambridge University Press, 1969); Arghiri Emmanuel, *Unequal Exchange* (New York: Modern Reader, 1972); , Samir Amin, *Unequal Development* (New York: Monthly Review Press, 1976); Immanuel Wallerstein, *The Modern World Sys-*

tem (New York: Academic Press, 1974); Peter B. Evans, *Dependent Development: The Alliance of Multinational, State and Local Capital in Brazil* (Princeton: Princeton University Press, 1979); and others.

47. I provide data in this chapter solely on the increasing numbers of transnational nongovernmental advocacy organizations. This classification is similar to the one offered in Khagram, Riker, and Sikkink, 2002.

48. See Robert O. Keohane and Joseph S. Nye Jr. (1971); Peter Willets, ed., *Pressure Groups in the Global System: The Transnational Relations of Issue-Oriented Non-Governmental Organizations* (New York: St. Martin's Press, 1982); Jamie Leatherman et al., "International Institutions and Transnational Social Movement Organizations: Challenging the State in a Three-Level Game of Global Transformation" (paper delivered at International Studies Association, Washington, D.C., 28 March to 1 April 1994); and Paul K. Wapner, *Environmental Activism and World Civic Politics* (Albany: State University of New York Press, 1996).

49. See Risse-Kappen, "Introduction" (1995). Another set of data shows that while only 37 of these groups existed before 1875, by 1973 over 6,000 from the International Red Cross to the World Wildlife Fund for Nature were on record as having been founded.

50. These groups should be contrasted with those transnational actors clandestinely and often violently seeking to advance principled ideas (such as terrorists) or seeking to gain materially (organized criminals). See various issues of the journal *Transnational Organized Crime* (London: Frank Cass), and Bruce Hoffman, *Inside Terrorism* (New York: Columbia University Press, 1998).

51. See Kathryn Sikkink and Jackie Smith, "Infrastructures for Change: Transnational Organizations 1953–1993," in Khagram et al. (2002).

52. In fact, the establishment of transnational nongovernmental organizations often results from the activities of already existing transnational coalitions.

53. Identity groups and communities such as diasporas, migrant communities, and border groups which cross territorial boundaries in multiple ways are also transnational. However, these are not central to the analysis I offer in this study. See for example, Nina Glick Schiller, Linda Basch, and Christina Balnc-Szanton, eds., *Towards a Transnational Perspective on Migration: Race, Class, Ethnicity and Nationalism Reconsidered* (New York: New York Academy of Sciences, 1992), vol. 645; Arjun Appadurai, *Modernity at Large: Cultural Dimensions of Globalization* (Minneapolis: University of Minnesota Press, 1996); and Ulf Hannerz, *Transnational Connections: Cultures, People, Places* (London: Routledge, 1996). I also do not include trans-governmental relations in this list, because they do not include nongovernmental organizations. See Ann Marie Slaughter, "The Real New World Order," *Foreign Affairs* 76, no. 5 (September/October 1997).

54. Keck and Sikkink also add a common discourse to their definition of networks. Transnational networks of nongovernmental organizations are said to be based on shared normative concerns, as opposed to the instrumental goals of transnational corporations, or the common causal ideas of epistemic communities. See Margaret E. Keck and Kathryn Sikkink, *Activists Beyond Borders: Advocacy Networks in International Politics* (Ithaca: Cornell University Press, 1998); Peter B. Evans et al., eds., *States versus Markets in the World System* (Beverly Hills: Sage Publications, 1985); and Peter Haas, "Knowledge, Power, and International Policy Coordination," *International Organization* 46 (Winter 1992): 1–39. This means that all these actors can and do act strategically, but that they are motivated for different purposes or interests.

55. See Paul J. Nelson, *The World Bank and Non-Governmental Organizations: The Limits of Apolitical Development* (New York: St. Martin's Press, 1995).

56. On organizational fields see Walter W. Powell and Paul J. Dimaggio, *The New Institutionalism in Organizational Analysis* (Chicago: University of Chicago Press, 1991). For social fields, see Pierre Bourdieu, *The Logic of Practice* (Stanford: Stanford University Press, 1990).

57. Allison Brysk, "The International Politics of Indians in Latin America," *Latin American Perspectives* 23, no. 2, (1996), 38–57.

58. I define norms broadly as shared expectations of appropriateness held by a community of actors. Norms are institutions, but depending on the context, the term institution in this study is often used to denote more formalized procedures and structures that guide behavior. Thus the notion that environmental factors must be included in the evaluation of development projects is a norm, while the more explicit requirement that environmental impact assessments must be completed before development projects can be implemented would be considered an institution. For discussions on norms and rules in social life from differing perspectives see, for example, Jon Elster, *Nuts and Bolts for the Social Sciences* (Cambridge: Cambridge University Press, 1989); Anthony Giddens, *The Constitution of Society* (Berkeley: University of California Press, 1984); John Searle, *Speech Acts: An Essay in the Philosophy of Language* (Cambridge: Cambridge University Press, 1969); and Fred R. Dallmayr and Thomas A. McCarthy, eds., *Understanding and Social Inquiry* (Notre Dame: University of Notre Dame Press, 1977). For discussions of norms in world affairs, see Martha Finnemore, *National Interests in International Society* (Ithaca: Cornell University Press, 1996); Peter Katzenstein, *The Culture of National Security: Norms and Identity in World Politics* (New York: Columbia University Press, 1996); and Martha Finnemore and Kathryn Sikkink, "International Norm Dynamics and Political Change," *International Organization* 52, no. 4 (1998).

59. Scholars have debated and remain divided over the existence and efficacy of norms and institutions in world politics. See Stephen D. Krasner, *Structural Conflict: The Third World Against Global Liberalism* (Berkeley: University of California Press, 1985); Robert O. Keohane, "International Institutions: Two Approaches," *International Studies Quarterly* 32 (1988): 379–396; and Volker Rittberger, *Regime Theory and International Relations* (Oxford: Oxford University Press, 1993). The constructivist challenge in international relations has involved a strong claim that norms, identities, and culture not only influence the dynamics of world politics, but also that the core variables of states, power capabilities, and interests are reciprocally constituted and transformed by these broader sets of ideas and institutions. See Alexander Wendt and Raymond Duvall, "Institutions and Order in the International System," in *Global Changes and Theoretical Challenges*, ed. Ernst Otto Czempiel and James Rosenau (London: Lexington, 1989); Alexander Wendt, "The Agent-Structure Problem in International Relations Theory," *International Organization* 41 (1987): 335–370, and "Anarchy Is What States Make of It: The Social Construction of Power Politics," *International Organization* 46 (1992): 391–425; David Dessler, "What's at Stake in the Agent-Structure Debate?" *International Organization* 43, no. 3 (Summer 1989); Friedrich Kratochwil and John Ruggie, "International Organization: A State of the Art or an Art of the State," *International Organization* 40 (1986): 753; John G. Ruggie, "Transactors and Change: Embedded Liberalism in a Post-War Economic Order," in *International Regimes*, ed. Stephen D. Krasner (Ithaca: Cornell University Press, 1983); John G. Ruggie, *Multilateralism Matters: The Theory and Praxis of an Institutional Form* (New

York: Columbia University Press, 1993); and Yosef Lapid and Friedrich Kratochwil, *The Return of Culture and Identity in IR Theory* (Boulder, Colo.: Lynne Rienner, 1996). For more empirically grounded accounts, see Ethan A. Nadleman, "Global Prohibition Regimes: The Evolution of Norms in International Society," *International Organization* 44 (1990): 479–526; Jutta Weldes, "Constructing National Interests: The Logic of U.S. National Security in the Postwar Era" (Ph.D. diss., University of Minnesota, 1992); Richard Price, "Genealogy of the Chemical Weapons Taboo," *International Organization* 49 (1995); Audie Klotz, *Norms in International Relations: The Struggle Against Apartheid* (Ithaca: Cornell University Press, 1995); Finnemore (1996); and Katzenstein (1996).

60. See George M. Thomas et al., *Institutional Structure: Constituting State, Society, and the Individual* (Newbury Park: Sage, 1987); and John Meyer et al., "World Society and the Nation State," *American Journal of Sociology* 103 (1997): 144–181. For a similar but more interpretative research program along these lines, see Barry Buzan, "From International System to International Society: Structural Realism and Regime Theory Meet the English School," *International Organization* 47, no. 3 (Summer 1993): 327–352. For a classic work from this perspective, see Hedley Bull and Adam Watson, eds., *The Expansion of International Society* (Oxford: Clarendon Press, 1984).

61. Regulatory norms define standards of appropriate behavior, constitutive norms define actor identities, practical norms focus on commonly accepted notions of best solutions, and evaluative norms stress questions of morality. See Katzenstein (1996).

62. The existence of norms that legitimate the state as the only natural form of political organization in the world system provides a powerful explanation for why completely dysfunctional states not only persist, but also why these entities continue to be propped up by other states and international organizations. But the same principles of sovereignty and self-determination can be used by groups to de-privilege states in their demands to secede. See Robert Jackson and Carl Rosberg, "Why Africa's Weak States Persist: Empirical and Juridical Statehood," *World Politics* 35 (1982): 1–24; John W. Meyer, "The World Polity and the Authority of the Nation-State," in Thomas et al. (1987); David Strang, "From Dependency to Sovereignty: An Event History Analysis of Decolonization," *American Sociological Review* 55 (1990): 846–860; and Cynthia Weber, *Simulating Sovereignty: Intervention, the State, and Symbolic Exchange* (Cambridge: Cambridge University Press, 1995).

63. National security has perhaps been the central role ascribed to all states and state leaders.

64. See Dana P. Eyre and Marc C. Suchman, "Military Procurement as Rational Myth: Notes on the Social Construction of Weapons Proliferation," *Sociological Forum* 7 (1992): 137–161; Gili Drori, "Global Discourse, State Policy, and National Governance: The Case of the Globalization of Science Policy," *MacArthur Foundation's Consortium—Research Series on International Peace and Cooperation*, Working Paper no. 7 (1997); Francisco O. Ramirez and Jane Wise, "The Political Incorporation of Women," in *National Development and the World System*, ed. John Meyer and Michael Hannan (Chicago: University of Chicago Press, 1979); and John Boli, "The Expansion of Nation-States," in Thomas et al. (1987).

65. John Meyer et al., "The Rise of an Environmental Sector in World Society" (paper presented at the Annual Meetings of the American Sociological Association, Los Angeles, 1994), 15.

66. See David John Frank et al., "The Nation-State and the Natural Environment over the Twentieth Century," *American Sociological Review* (2000): 96–116.

67. See Oran R. Young, "The Politics of International Regime Formation: Managing Natural Resources and the Environment," *International Organization* 43, no. 3 (Summer 1989): 349–375; and Oran R. Young, *The Effectiveness of International Environmental Regimes: Causal Connections and Behaviorial Mechanisms* (Cambridge: MIT Press, 1999).

68. The report of the preparatory meeting for the UN Stockholm Conference on the Human Environment in September 1971 stated that while development was the cure for environmental problems—like those that arise from widespread poverty—in developing countries, development had produced environmental problems in industrialized countries. Correspondingly, the report stated that "as the process of development gets under way, the latter type of problem is likely to assume greater importance. The process of agricultural growth and transformation, for example, will involve the construction of reservoirs and irrigation systems, the clearing of forests, the use of fertilizers and pesticides, and the establishment of new communities. These processes will certainly have environmental implications. . . ." Quote from Darryl D'Monte, *Temples or Tombs?, Industry Versus Environment: Three Controversies* (New Delhi: Centre for Science and Environment, 1985), 8.

69. See various chapters in Thomas et al. (1987).

70. See Samuel P. Huntington, *The Third Wave: Democratization in the Late Twentieth Century* (Norman: University of Oklahoma Press, 1991); Francisco O. Ramirez and John W. Meyer, *Citizenship Principles, Human Rights, and the National Incorporation of Women, 1870–1990*, proposal to U.S. National Science Foundation, 1996; Karen Brown Thompson, "Changing Global Norms Concerning Women's and Children's Rights and their Implications for State-Citizen Relations," in Riker, Khagram, and Sikkink (2002); and Connie McNeely, *Constructing the Nation-State: International Organization and Prescriptive Action* (Westport, Conn.: Greenwood Press, 1995).

71. See Human Rights Watch and Natural Resources Defense Council, *Defending the Earth, Abuses of Human Rights and the Environment* (New York: Human Rights Watch and Natural Resources Defense Council, 1992); and Aaron Sachs, "Human Rights and the Environment," in Brown et al. (1996). See also campaigns of Earthrights International.

72. See John P. Humphrey, *Human Rights and the United Nations: A Great Adventure* (Dobbs Ferry, N.Y.: Transnational, 1984).

73. As noted before, there is an interactive process by which transnational norms are reshaped by local dynamics. While the origins of many 20th-century globalizing norms have been western, and although many of these are still likely to arise in the West, the dynamics of constructing a world society have become much more multidirectional and interactive.

74. Similar arguments and evidence could be offered in the cases of indigenous peoples, gender justice, children's rights, peace, and a range of other areas. See David P. Forsythe, *Human Rights and World Politics*, 2d ed. (Lincoln: University of Nebraska Press, 1989); N. G. Onuf and V. Spike Peterson, "Human Rights from an International Regime Perspective," *Journal of International Affairs* 38 (Winter 1984): 329–342; Jack Donnelly, "International Human Rights: A Regime Analysis," *International Organization* 40, no. 3 (Summer 1986): 599–642; and Jack Donnelly, *Universal Human Rights in Theory and Practice* (Ithaca: Cornell University Press, 1989).

75. For a treatment of causal mechanisms in the social sciences, see Jon Elster (1989).

76. The tremendous number of human rights violations that have been and continue to be perpetrated is ample evidence of this reality.

77. See Finnemore (1996) and Stephen D. Krasner, *Sovereignty: Organized Hypocrisy* (Princeton: Princeton University Press, 1999).

78. The rapid formal embracing of neoliberal procedures and structures in the third world with uneven implementation in practice is an excellent example of this. For the importance of domestic structures in mediating the ultimate outcomes from an earlier global spread of economic discourse, see Peter A. Hall, ed., *The Political Power of Economic Ideas: Keynesianism across Nations* (Princeton: Princeton University Press, 1989).

79. See Peter L. Berger and Thomas Luckmann, *The Social Construction of Reality: A Treatise in the Sociology of Knowledge* (Garden City: Doubleday, 1966); Pierre Bourdieu, *The Logic of Practice* (Stanford: Stanford University Press, 1980); and Anthony Giddens, *The Constitution of Society: Outline of the Theory of Structuration* (Cambridge: Polity Press, 1984).

80. On causal pathways, see David Dessler, "The Architecture of Causal Analysis" (working paper, Center for International Affairs, Harvard University, April 1992).

81. See Humphrey (1984).

82. See Freedom House, *Freedom in the World,* various years.

83. See Ans Kolk et al., "International Codes of Conduct and Corporate Social Responsibility," *Transnational Corporations* 8, no. 1 (April 1999).

84. See Haas (1992), 1–39; McNeely (1995), and Martha Finnemore, "Norms, Culture and World Politics: Insights from Sociology's Institutionalism," *International Organization* 49, no. 3 (1995): 325–348.

85. See Terry Lynn Karl and Philippe C. Schmitter, "Democratization Around the Globe: Opportunities and Risks," in *World Security: Challenges for a New Century,* ed. Michael T. Klare and Daniel C. Thomas (New York: St. Martins Press, 1992), 43–62; Kathryn Sikkink (1993); Allison Brysk (1993), 259–285; Chilton (1994), 151–181; and Jackie Smith et al, "Globalizing Human Rights: The Work of Transnational Human Rights NGOs in the 1990s," *Human Rights Quarterly* 20 (1998): 379–412).

86. Krasner (1983), 2.

87. Many international organizations and multinational corporations are also resistant to changing actual behavior, but the analysis of these actors is somewhat different from those of states.

88. See Barrington Moore Jr., *Social Origins of Dictatorship and Democracy* (Boston: Beacon Press, 1966); Samuel P. Huntington, *Political Order in Changing Societies* (New Haven: Yale University Press, 1968); and Robert H. Bates, *Markets and States in Tropical Africa: The Political Basis of Agricultural Policies* (Berkeley: University of California Press, 1981). The proposition that states are shaped by external factors does not preclude but complements domestic level accounts. See also Fernando Henrique Cardoso and Enzo Faletto, *Dependencia y Desarrollo en America Latina; Ensayo de Interpretacion Sociologica* (Mexico: Siglo Veintiuno Editores, 1969); Guillermo A. O'Donnell, *Modernization and Bureaucratic-authorities; Studies in South American Politics,* Politics of Modernization Series, no. 9 (Berkeley, Institute of International Studies, University of California, 1973); Peter B. Evans (1979); Robert D. Putnam, "Diplomacy and Domestic Politics: The Logic of Two-Level Games," *International Organization* (Summer 1988): 427–460; and Peter B. Evans et al., eds., *Double-edged Diplomacy: International Bargaining and Domestic Politics* (Berkeley: University of California Press, 1993).

89. Regime in the domestic arena denotes the set of political and administrative institutions that determine: (1) the forms and channels of access to principle state positions;

(2) the characteristics of actors who are admitted to and excluded from such access; (3) the tactics that these actors can use to gain access, and (4) the procedures that are acceptable in the formulation and implementation of state decisions. Democracy, then, is a regime in which there exists at a minimum: (1) meaningful competition for all effective state positions through regular, free, and fair elections; (2) a highly inclusive level of political participation in which no social group is prevented from exercising the rights of citizenship; and (3) a level of civil and political liberties secured through equality under the rule of law. See Larry Diamond et al., chap. 1 in *Politics in Developing Countries: Comparing Experiences with Democracy* (Boulder: Lynne Rienner, 1995). For a more extensive definition of political democracy with which I am sympathetic, see Philippe C. Schmitter and Terry Lynn Karl, "What Democracy Is . . . and Is Not," *Journal of Democracy* 2, no. 3 (Summer 1991): 75–88. For an excellent analysis of why the concepts of bureaucracy, regime, and government should be kept analytically distinct when examining state institutions, see Stephanie Lawson, "Conceptual Issues in the Comparative Study of Regime Change and Democratization," *Comparative Politics* (January 1993): 183–205.

90. See Ted Gurr, "On the Political Consequences of Scarcity and Economic Decline," *International Studies Quarterly* 29, no. 1 (March 1985): 51–75.

91. I do not elaborate in this section a major current of social movement theorizing that focuses less on external resource conflicts and macro-political opportunity structures and more on internal, micro-mobilizational strategies, discourses and identity dynamics. But the historical and ethnographic analysis presented in subsequent chapters offers a glimpse into these latter elements of collective organization and mobilization. For key texts from this "new social movement" approach see Clauss Offe, "New Social Movements: Changing Boundaries of the Political," *Social Research* 52 (1985): 817–868; Russell J. Dalton and Manfred Kuechler, eds., *Challenging the Political Order: New Social and Political Movements in Western Democracies* (New York: Oxford University Press, 1990); and Arturo Escobar and Sonia E. Alvarez, eds., *The Making of Social Movements in Latin America: Identity, Strategy and Democracy* (Boulder, Colo.: Westview, 1992). For a pathbreaking attempt to "more successfully read, interpret, and understand the often fugitive political conduct of subordinate groups" see James Scott, *Domination and the Arts of Resistance* (New Haven: Yale University Press, 1990).

92. On the other hand, they are also strengthened by their transnational linkages to foreign and transnational nongovernmental organizations.

93. In the environmental issue area, see P. J. Sands, "The Role of Non-Governmental Organizations in Enforcing International Environmental Law," in *Control over Compliance With International Law*, ed. William E. Butler (The Hague, Netherlands: Martinus Nijhoff, 1991), 61–68; Oran Young, *International Cooperation: Building Regimes for Natural Resources and the Environment* (Ithaca: Cornell University Press, 1993); and Oran Young, *Global Environmental Change and International Governance* (Hanover, N.H.: Dartmouth Press, 1996).

94. See Charles Tilley, *From Mobilization to Revolution* (Englewood Cliffs: Prentice Hall, 1978); Sidney Tarrow, "National Politics and Collective Action: Recent Theory and Research in Western Europe and the United States," *Annual Review of Sociology* 14 (1988): 421–440; Doug McAdam et al., *Comparative Perspectives on Social Movements* (Cambridge: Cambridge University Press, 1996); and Doug McAdam and David A Snow, eds., *Social Movements: Readings on Their Emergence, Mobilization and Dynamics* (Los Angeles: Roxbury Publishing, 1997).

95. See Sidney Tarrow, *Power in Movement* (New York and Cambridge: Cambridge University Press, 1994), Betty H. Zisk, *The Politics of Transformation: Local Activism in the Peace and Environmental Movements* (Westport: Praeger, 1992); and Margaret E. Keck, "Social Equity and Environmental Politics in Brazil: Lessons From the Rubber Tapper of Acre," *Comparative Politics* 27, no. 4 (July 1995): 409–424.

96. See William Ophuls, *Ecology and the Politics of Scarcity: Prologue to a Political Theory of the Steady State* (San Francisco: W. H. Freeman, 1977); Huntington (1968),; and Stephen Haggard, *Pathways from the Periphery: The Politics of Growth in Newly Industrializing Countries* (Ithaca: Cornell University Press, 1990).

97. Adam Przeworski and Fernando Limongi, "Political Regimes and Economic Growth," *Journal of Economic Perspectives* 7, no. 3 (1993): 54.

98. Haggard (1990): 262.

99. See Pranab Bardhan, *Development and Change* (Bombay: Oxford, 1993), 633–639. Also see Peter Evans, "Predatory, Developmental, and Other Apparatuses: A Comparative Political Economy Perspective on the Third World State," *Sociological Forum* 4, no. 4 (December 1989): 561–587.

100. See Amartya Kumar Sen, *Poverty and Famines* (Oxford: Clarendon Press, 1981).

101. See Douglass North, *Institutions, Institutional Change, and Economic Performance* (Cambridge: Cambridge University Press, 1990).

102. See Adam Przeworski, *Democracy and the Market* (Cambridge: Cambridge University Press, 1991).

103. See George Psacharopoulos, *Economics of Education* (Oxford: Pergamon Press, 1985); and Paul Schultz, *State of Development Economics* (Oxford: Basil Blackwell, 1988).

104. See Adam Przeworski and Fernando Limongi, "Democracy and Development," paper presented to the Nobel Symposium on "Democracy's Victory and Crisis," Uppsala University, Sweden, 27–30 August 1994: 13. See also Adam Przeworski and Fernando Limongi, *Democracy and Development in South America, 1946–1988* (Madrid, Spain: Instituto Juan March de Estudios e Investigaciones, Centre de Estios Avanzados en Ciencias Sociales, 1994).

105. An expanded view of development was adopted by the *Human Development Report* of 1990 produced by the United Nations Development Program. For an exploration of the implications of taking such a view on development policy, see Sudhir Anand and Martin Ravallion, "Human Development in Poor Countries: On the Role of Private Incomes and Public Services," *Journal of Economic Perspectives* 7, no. 1 (Winter 1993): 133–150.

106. See Bardhan (1993).

107. See Miles D. Wolpin, *Militarization, Internal Repression, and Social Welfare in the Third World* (New York: St. Martin's Press, 1986).

108. See Adam Przeworski et al., *Democracy and Development: Political Institutions and Well Being in the World, 1950–1990* (Cambridge: Cambridge University Press, 2000).

109. See Diamond et al. (1995).

110. See Paul Ehrlich, *The Population Explosion* (New York: Simon and Schuster, 1990).

111. See Paul Ehrlich, *The Population Bomb* (New York: Ballantine, 1968). The following section has benefited from Rodger A. Payne, "Freedom and the Environment," *Journal of Democracy* 6, no. 3 (July 1995): 41–53. I do not share his inclusion of market-oriented economies in the definition of democracy. Payne also does not highlight the existence and functioning of judicial systems and the courts as a critical institutional factor making democracies more conducive to environmentally sustainable practices.

112. See Robert Paehlke, *Environmentalism and the Future of Progressive Politics* (New Haven: Yale University Press, 1989); Walter Truett Andersen, ed., *Rethinking Liberalism* (New York: Avon, 1983); and Mark E. Kann, "Beyond Environmentalism: The Biological Foundations of Governance," in *Controversies in Environmental Policy,* ed. Sheldon Kamiecki, Robert O'Brien, and Michael Clarke (Albany: State University of New York Press, 1986).

113. See Michael Waller and Frances Millard, "Environmental Politics in Eastern Europe," *Environmental Politics* 1 (Summer 1992): 159–185; Sheldon Kamienicki, "Political Mobilization, Agenda Building and International Environmental Policy," *Journal of International Affairs* 44 (Winter 1991), Charles E. Ziegler, *Environmental Policy in the USSR* (Amherst: University of Massachusetts, 1987); and Philip R. Pryde, *Environmental Management in the Soviet Union* (New York: Cambridge University Press, 1991).

114. See David Vogel and Veronica Kun, "The Comparative Study of Environmental Policy: A Review of the Literature," in *Comparative Policy Research: Learning From Experience,* ed. Meinholf Dierkes, Hans N. Weiler, and Ariane Berthoin Antal (New York: St. Martin's Press, 1989); and Sheldon Kamienicki and Eliz Samasarian, "Conducting Comparative Research on Environmental Policy," *Natural Resources Journal* 30 (Spring 1990).

115. See Barbara Jancar, *Environmental Management in the Soviet Union and Yugoslavia: Structure and Regulation in Federal Communist States* (Durham: Duke University Press, 1987); and Barbara Jancar, "Democracy and the Environment in Eastern Europe and the Soviet Union," *Harvard International Review* 12 (Summer 1990).

116. Authoritarian regimes are also likely to generate higher levels of corruption, and levels of corruption are significantly correlated with environmental damage, human rights abuses, and unsustainable human development more broadly.

117. See Robert Dahl, *Polyarchy* (New Haven: Yale University Press, 1971); and Diamond et al. (1995).

118. It may be that different *types* of democracies or *degrees* of democratization are more or less conducive to sustainable development because they offer more access and support to this type of domestic and transnational political action.

119. The promise is that as environmental problems become more and more of a priority to citizens, in large part due to the activities of advocacy organizations; these will condition changes in the platforms and practices of powerful non-state actors, politicians, political parties, and state authorities.

120. See Ferdinand Muller-Rommel, ed., *New Politics in Western Europe: The Rise and Success of Green Parties* (Boulder, Colo.: Westview, 1989); Wolfgang Rudig, *The Green Wave: A Comparative Analysis of Ecological Parties* (Oxford: Polity, 1993); and Dick Richardson, *The Green Challenge* (London: Routledge, 1995).

121. See Marian A. L. Miller, *The Third World in Global Environmental Politics* (Boulder, Colo.: Lynne Rienner, 1995).

122. The term state is being used, depending on the context, as an actor or as a set of institutions, although it is better to argue that governments and government officials are the actors who deploy state authority. Following Alfred Stepan and others, the state as a set of institutions is generally defined as "the continuous administrative, legal, bureaucratic, and coercive systems that attempt not only to structure relationships between civil society and public authority in a polity but also to structure many crucial relationships within civil society as well." See his *The State and Society: Peru in Comparative Perspective* (Princeton: Princeton University Press, 1978). See also Atul Kohli, ed., *The State and*

Development in the Third World (Princeton: Princeton University Press, 1986); Donald Rothchild and Naomi Chazan, *State and Society in Africa* (Boulder, Colo.: Westview, 1988); Peter Katzenstein, *Small States in the World Economy* (Ithaca: Cornell University Press, 1985); and Terry Lynn Karl, *The Paradox of Plenty* (Berkeley: University of California Press, 1998).

123. See Skocpol et al., eds., *Bringing the State Back In;* Kohli (1986); Risse-Kappen, "Introduction" (1995); and Peter B. Evans, *Embedded Autonomy: States and Industrial Transformation* (Princeton: Princeton University Press, 1995).

124. Risse-Kappen, "Introduction" (1995).

125. See Stephen D. Krasner, "Realism and Transnationalism"; and Thomas Risse-Kappen, "Introduction," both in Risse Kappen (1995).

126. See Karl and Schmitter in Klare and Thomas (1992), 43–62; and Chilton (1994), 151–181.

127. See Adam Przeworski and Henry Teune, *The Logic of Comparative Social Inquiry* (Malabar: R. E. Krieger, 1982). In addition to these methods, a variant of multi-sited ethnography was conducted.

128. See Alexander George and Andrew Bennett, *Case Studies and Theory Development* (Cambridge: MIT Press, 2002).

129. See Henry E. Brady and David Collier, eds., *Rethinking Social Inquiry: Diverse Tools, Shared Standards,* New York: Rowman & Littlefield, 2002).

130. I also conducted participant observation in several settings, including as Senior Advisor for Policy and Institutional Analysis at the World Commission on Dams from 1998–2000. The broader research strategy was one of systematic triangulation of various units, research methods, types of data, and social science philosophies. See Todd Jick, "Mixing Qualitative and Quantitative Methods: Triangulation in Action," *Administrative Science Quarterly* 24, no. 4 (1999): 602–611; Ann Chih Lin, "Bridging Positivist And Interpretivist Approaches to Qualitative Methods," *Policy Studies Journal* 26, no. 1 (1998): 162–180; and James Mahoney, "Nominal, Ordinal, and Narrative Appraisal in Macro-Causal Analysis," *American Journal of Sociology* 104, no. 4 (1999): 1154–1196.

2. DAMS, DEMOCRACY, AND DEVELOPMENT IN INDIA

1. C. V. J. Varma and K. R. Saxena, eds., *Modern Temples of India: Selected Speeches of Jawaharlal Nehru at Irrigation and Power Projects and Various Technical Meetings of Engineers and Scientists* (New Delhi: Central Board of Irrigation and Power, April 1989), 17.

2. Interview, Barwani, Madhya Pradesh, 14 May 1993.

3. This phrase was popularized by Darryl D'Monte, *Temples or Tombs—Industry versus Environment: Three Controversies* (New Delhi: Centre for Science and Environment, 1985).

4. Enakshi Ganguly Thukral, ed., *Big Dams, Displaced People: Rivers of Sorrow, Rivers of Change* (New Delhi: Sage Publications, 1992).

5. B. D. Dhawan, ed., *Big Dams: Claims, Counter-Claims* (New Delhi: Commonwealth Publishers, 1990).

6. See Planning Commission, *First Five Year Plan: Irrigation and Power* (New Delhi: Government of India, 1954).

7. The demobilization of civil society in India subsequent to the transfer of power re-

sembles, quite strikingly, the same pattern generally found during post-transition phases to democracy during the fourth wave of democratization such as in South Africa after 1994. For this argument, see Guillermo O'Donnell and Philippe Schmitter, *Transitions from Authoritarian Rule: Tentative Conclusions about Uncertain Democracies* (Baltimore: Johns Hopkins University Press, 1987). For the most comprehensive account of India's political economy of development during this early period, see Francine R. Frankel, *India's Political Economy, 1947–77: The Gradual Revolution* (Princeton: Princeton University Press, 1978). For an analysis of the India's "steel frame bureaucracy" inherited from the British, see David C. Potter, *India's Political Administrators: From ICS to IAS* (Delhi: Oxford University Press, 1996).

8. For example, India's First Five Year Plan states that "cheap electric power is essential for the development of a country. In fact, modern life depends so largely on the use of electricity that the quantity of electricity use per capita in a country is an index of its material development and of the standard of living attained in it." Planning Commission (1954), 345. See also Sukhamoy Chakravarty, *Development Planning: The Indian Experience* (Oxford: Oxford University Press, 1987).

9. For differences between Nehru and Gandhi, see Raj Krishna, "Nehru-Gandhi Polarity and Economic Policy" (lecture at Nehru Memorial Museum Library, 15 November 1977). For a historical analysis of the adoption of a Nehruvian development strategy at Independence, see Frankel (1978), chaps. 1–3.

10. For an account that corroborates this view, see Ashutosh Varshney, *Democracy, Development, and the Countryside: Urban-Rural Struggles in India* (Cambridge: Cambridge University Press, 1995), chap. 2.

11. These three groups constitute what Pranab Bardhan has termed India's dominant coalition of proprietary classes. See Pranab Bardhan, *The Political Economy of Development in India* (Delhi: Oxford University Press, 1984), chaps. 6–9.

12. The hypothesis that authorities use development projects for patronage purposes is best elaborated by Robert H. Bates in his *Markets and States in Tropical Africa: The Political Basis of Agricultural Policies* (Berkeley: University of California Press, 1981). See also Pradip Kumar Bose, "Political Economy of Irrigation: A Note," (Surat, India: Centre for Social Studies 1987), unpublished manuscript.

13. For the general role of World Bank assistance to India, the country that has been the largest recipient of World Bank aid, see C. P. Bhambhri, *The World Bank and India* (New Delhi: Vikas, 1980). For a critical view of the World Bank's role in supporting big dam building in the third world, see Leonard Sklar and Patrick McCully, "Damming the Rivers: The World Bank's Lending for Large Dams," *International Rivers Network Paper Number 5* (Berkeley: IRN, 1994). For an analysis of the involvement of the United States in India's river valley development programs, see Henry Hart, *Administrative Aspects of River Valley Development* (Bombay: Asia Publishing House, 1961) and for an analysis of how India increasingly shifted its ties away from the United States, see Sanjoy Basu, "Nonalignment and Economic Development: Indian State Strategies, 1947–1962," in *States versus Markets in the World-System*, ed. Peter Evans et al. (Beverly Hills: Sage Publications, 1985), chap. 6.

14. For a celebratory analysis of the Bhakra Nangal and other major river valley projects in India, see B. C. Verghese, *Winning the Future: From Bhakra to Narmada, Tehri, Rajasthan Canal* (New Delhi: Konark, 1994).

15. This account is based on the following sources: Philip Viegas, "The Hirakud Dam

Oustees: Thirty Years After," in Thakral (1992), 29–53; and S. K. Pattanaik, B. Das, and A. Mishra, "Hirakud Dam Project: Expectations and Realities," in *People and Dams* (New Delhi: PRIA, 1987), 47–59. PRIA is an acronym for The Society for Participatory Research in Asia.

16. As Paul Brass notes, "two strict rules have been followed with dissident domestic ethnic, religious, linguistic and cultural group demands. The first is that no secessionist movement will be entertained and that any group which takes up a secessionist stance will, while it is weak, be ignored and treated as illegitimate, but, should it develop significant strength, be smashed, with armed force if necessary," in *The Politics of India Since Independence* (Cambridge: Cambridge University Press, 1994), 7.

17. Government of Orissa, *Report on the Benefits of Hirakud Irrigation: A Socio-Economic Study* (Cuttack, Orissa: Bureau of Statistics and Economics, 1968), 11.

18. Prakash Sawant, *River Dam Construction and Resettlement of Affected Villages* (New Delhi: Inter-India Publishers, 1990).

19. *The Bombay Chronicle,* 12 April 1948, 5.

20. Varma and Saxena, 1989.

21. See International Commission on Large Dams, *World Register of Dams* (Paris: International Commission on Large Dams, 1998).

22. Raj Krishna, "Stagnant Parameters," *Seminar,* New Delhi, January 1984. Partially due to the vast numbers of incomplete big dam projects, the share of electricity supplied by hydropower in India fell from over fifty to about twenty percent during that same period.

23. Planning Commission: *Eighth Five Year Plan* (New Delhi: Government of India, 1992), 58.

24. On this issue, see the speech by Dr. C. C. Patel in *Symposium on Large Dams: Socio-Environmental and Techno-Economic Assessment* (New Delhi: Central Bureau of Irrigation and Power, December 1991).

25. For an overview of federalism and water resources in India, see Ramaswamy Iyer, "Indian Federalism and Water Resources," *Water Resources Development* 10, no. 2 (1994): 191–202.

26. State with a capital 'S' signifies the meso-level of India's federal system in this book. In India, the federal level is called the Centre.

27. For an analysis of this tension and its progression over the years, see Lloyd I. Rudolph and Suzanne Hoeber Rudolph, *In Pursuit of Lakshmi: The Political Economy of the Indian State* (Chicago: University of Chicago Press, 1987), chap. 3.

28. Ministry of Law, *The River Boards Act, 1956,* no. 49 (New Delhi: Government of India, 1956).

29. Ministry of Law, *The Inter-State Water Disputes Act, 1956,* no. 33 (New Delhi: Government of India, 1956).

30. See R. Sarkaria Commission, "Inter-State River Water Disputes," in *Report of the Commission on Centre-State Relations* (New Delhi: Government of India, 1988), 487–493.

31. On the financial dependence of States on the Centre, see P. R. Shukla and S. K. Roy Chowdhry, *Centre-State Finances in the Indian Economy* (New Delhi: Akashdeep, 1992).

32. Dr. A. N. Khosla, "Central Water and Power Commission," in *Twenty Five Years of the Central Water and Power Commission 1945–1970* (New Delhi: Government of India, Ministry of Irrigation and Power, 1971), 10.

33. For an analysis of relationship between political party dynamics and India's federal

system, see Mahendra Prasad Singh, "Political Parties and Political Economy of Federalism: A Paradigm Shift in Indian Politics," *Indian Journal of Social Science* 7, no. 2 (1994): 155–177.

34. International Commission on Large Dams, *World Register of Dams* (Paris: ICOLD, 1988); and *International Water Power and Dam Construction Handbook* (Paris: ICOLD, 1995).

35. India's Ministry of Water Resources estimated (optimistically) that 34 major irrigation projects would be completed by the end of the Eighth Plan in 1997. This target was not achieved. See Ministry of Water Resources, *Base Paper in Respect of Private Sector Participation in Irrigation and Multipurpose Projects* (New Delhi: Government of India, 5 July 1995).

36. Edward Goldsmith and Nicholas Hildyard, *The Social and Environmental Effects of Large Dams: A Review of the Literature,* vol. 3 (Cornwall, U.K.: Wadebridge Ecological Centre, 1992), vii.

37. International Commission on Large Dams, 1998. See also, for example, Ministry of Water Resources, *Base Paper in Respect of Private Sector Participation in Irrigation and Multipurpose Projects* (New Delhi: Government of India, 5 July 1995).

38. See Report of the Expert Committee, *Rise in the Costs of Irrigation and Multipurpose Projects* (New Delhi: Government of India, Ministry of Irrigation and Power, April 1973).

39. See budget allocations in the Fifth through Tenth Five Year Plans of the Government of India.

40. Less than ten years earlier, India's Finance Minister, Morarji Desai, speaking at a public meeting in the submergence zone of the very same project had stated, "We will request you to move from your houses after the dam comes up. If you move, it will be good, otherwise we shall release the waters and drown you all." Quoted in Patrick McCully, *Silenced Rivers: The Ecology and Politics of Large Dams* (London: Zed Books, 1996), 72.

41. Nevertheless, the Pong Dam Oustees Association continued to fight for the rights of the displaced throughout the following decade. See Renu Bhanot and Mridula Singh, "The Oustees of the Pong Dam: Their Search for a Home," in Thakral (1992), 101–102.

42. See Enakshi Ganguly Thakral, "Introduction," in Thakral (1992), 9.

43. As Akileshwar Pathak correctly notes, "The globalisation (sic) of environmental considerations by the late sixties shifted the environmental agenda from the concerns of pollution in the developed countries to natural resources degradation in the developing countries." See his *Contested Domains: The State, Peasants, and Forests in Contemporary India* (New Delhi: Sage Publications, 1994), 32. See also Akhil Gupta, "Peasants and Global Environmentalism: Safeguarding the Future of 'Our World' or Initiating a New Form of Governmentality?" in *Postcolonial Developments: Agriculture in the Making of Modern India* (Durham, N.C.: Duke University Press, 1998), concluding chapter.

44. Government of India, *The Fourth Five Year Plan* (New Delhi: Government of India Press, 1970).

45. See Shekhar Singh, ed., "Introduction," *Environmental Policy in India* (New Delhi: Indian Institute of Public Administration, 1984).

46. Government of India, *Tiwari Committee Report on the State of India's Environment* (New Delhi: Government of India Press, 1980). The environment went on to become a central issue in India's development debate during the 1980s. See, for example, Anil Agarwal, *The State of India's Environment* (New Delhi: Centre for Science and Environment, 1982).

47. This section is greatly indebted to Darryl D'Monte, 1985.

48. Jayal would later become co-director of the Indian National Trust for Art and Cultural Heritage (INTACH), a Delhi-based nongovernmental organization that subsequently played an important role in anti-dam campaigns and the broader anti-dam movement in India.

49. See NCEPC, *Report of the Task Force for the Ecologically Planning of the Western Ghats* (New Delhi: Government of India, 1977).

50. IUCN's membership includes states and private sector companies, which makes it somewhat different from most transnational nongovernmental advocacy organizations. However, IUCN is a nonprofit group that nonviolently and publicly promotes social change for environmental sustainable development.

51. The transnational campaign also involved an attempt to save the lion-tailed monkey, a species only found in the Silent Valley whose survival was greatly endangered by the Silent Valley Project. See Steven Green and Karen Markowski, *Primate Conservation* (New York: Academic Press, 1977).

52. See M. K. Prasad et al., *The Silent Valley Hydro-Electric Project: A Techno-Economic and Socio-Political Assessment* (Trivandaram, Kerala: KSSP Health and Environment Brigade, 1979).

53. Ministry of Agriculture, *Development of the Silent Valley Reserve Forests, Kerala* (New Delhi: Government of India, 1979).

54. See C. K. Kochukosky, "Silent Valley: Fact or Fiction" (unpublished manuscript, 1980).

55. See KSEB, *Silent Valley Project and Silent Valley: An Assessment of the Controversy* (Trivandarum, Government of Kerala, 1980).

56. *Report of the Joint Committee on the Silent Valley* (New Delhi: Government of India and Government of Kerala, April 1983).

57. See Dr. R. N. Raj, "Is It Worthwhile to Ruin Forest Wealth for Electricity?" *Free Press,* Indore, Madhya Pradesh, 2 December 1987.

58. Anon., "Centre Urged to Clear Bodhghat Project," *Hindu,* New Delhi, 26 October 1987.

59. Quote in Darryl D'Monte (1985), 58.

60. Personal interviews, Bastar District, Madhya Pradesh, 24 February 1995.

61. See editorial, "Dithering on Project," *Hindustan Times,* New Delhi, 12 March 1987.

62. See International Rivers Network, "Bank Halts Bodhghat Funding," *World Rivers Review* (September–October 1988).

63. See Anon., "India Province Cancels Project," *Hydro Review Worldwide* (Summer 1995).

64. This section has benefited from Bruce Rich, *Mortgaging the Earth: The World Bank, Environmental Impoverishment, and the Crisis of Development* (Boston: Beacon, 1994), 44–46.

65. See Kavaljit Singh, "The Chandil Nightmare" and "Subarnarekha Multipurpose Project: For Whose Benefit?" (unpublished, New Delhi, Public Interest Research Group, n.d.).

66. See Pramod Parajuli, "World Bank in the Defensive: Adivasis Have a Better Cause to Displace the Subarnarekha Dam in Singhbhum, Bihar" (unpublished manuscript, Stanford University. 22 August 1988).

67. World Bank, India-Subarnarekha Irrigation Project (Cr. 1289–IN)—Resettlement Emergency Action. Office Memorandum, 16 June 1988.

68. Kavaljit Singh, Anil Singh, et al., "Police Arrest Peaceful Demonstrators" (unpublished statement signed by ten Indian Nongovernmental Organizations, New Delhi, 6 May 1991).

69. Clarence Mahoney, "Environmental and Project Displacement of Population in India, Part I: Development and Deracination," *Universities Field Staff Report*, no. 14 (Indianapolis: University Field Staff International, 1990–91).

70. Lori Udall et al., Letter to Heinz Virgin, Director, India Department, World Bank, on the Subernarekha Project in India (Washington, D.C.: Environmental Defense Fund, May 16, 1991). Shortly before that, on 2 May 1991, Peggy Hallward of Probe International in Canada sent a letter critical of the Subarnarekha Project to Mr. Frank Potter, Executive Director of the World Bank.

71. Senate Committee on Appropriations, Subcommittee on Foreign Operations, *Statement of Bruce M. Rich on behalf of Environmental Defense Fund et al.*, 102d Congress, 1st sess., 25 June 1991.

72. Probe International, "Subernarekha Project in India: Uproots Tribal People, Transforms River Basin," press release, Toronto (September 1991).

73. The following account is based partially on Verghese (1994), 75–117; The Indian National Trust for Arts and Cultural Heritage (INTACH), *The Tehri Dam: A Prescription for Disaster* (New Delhi: INTACH, 1987); and numerous interviews with participants in the Tehri controversy.

74. See Tehri Action Group, "Tehri Dam: An Unsafe and Non-Viable Project" (paper presented in New Delhi, June 1996), 4. The paper was presented to the Prime Minister and Environment Minister.

75. The following account of the Chipko Movement is based on Akhileshwar Pathak, *Contested Domains: The State, Peasants and Forests in Contemporary India* (New Delhi: Sage Publications, 1994); Centre for Science and Environment, *The State of India's Environment: A Citizen's Report* (New Delhi: CSE, 1982); Thomas Weber, *Hugging the Trees: The Story of the Chipko Movement* (New Delhi: Viking, 1988); and Ramchandra Guha, *The Unquiet Woods: Ecological Change and Peasant Resistance in the Himalayas* (New Delhi: Oxford University Press, 1989).

76. The resistance by the women-folk of Reni also made the Chipko struggle a landmark event in the formation of the broader women's movement in India.

77. Quoted in Bharat Dogra, *The Debate on Large Dams* (New Delhi: Bharat Dogra, 1992), 67. See also all of chap. 12, "A Detailed Report on the Tehri Dam Project," 64–99.

78. Letter from S. K. Roy, Chairman of the Working Group for the Assessment of the Environmental Impact of the Tehri Dam, to T. N. Seshan, Secretary of the Department of Forests and Environment, 28 August 1986. The letter was published in INTACH (1987), 3–11.

79. Verghese (1994), 78.

80. See Sunderlal Bahugna, "Save the Mothers of Culture: Message for Tehri Struggle" (unpublished manuscript, 1997).

81. *Tehri Bandh Virodhi Sangarsh Samiti and Others vs. Respondents, Writ Petition 12829*, submitted to the Supreme Court of India, New Delhi, 1985.

82. N. D. Jayal and Divyang K. Chhya, *Written Submissions on Behalf of INTACH in the Supreme Court in Writ Petition 12829 of 1985* (New Delhi: INTACH, September 1986).

83. Justice K. N. Singh and Justice Kuldip Singh, *Supreme Court Order in Writ Petition 12829 of 1985* (New Delhi: Supreme Court of India, 7 November 1990).

84. Government of India, *Indo-Soviet Expert Negotiations on the Issue of Cooperation in Designing and Construction of the Rockfill Dam and Spillway of the Tehri-Hydro Power Complex, Protocol No. 3* (New Delhi: Government of India, 25 December 1988).

85. Indian National Trust for Arts and Cultural Heritage, *Bhagirath ki Pukar* (New Delhi: INTACH, February and August 1993).

86. Ministry of Environment and Forests, "Letter to Tehri Hydro Development Corporation: Environmental Clearance—Tehri Project." New Delhi, 19 July 1990.

87. Nalini D. Jayal and Shekhar Singh, *Writ Petition against the Tehri Dam Project.* Submitted to the Supreme Court of India, New Delhi, 12 December 1991. Interview with Shekhar Singh, New Delhi, 14 December 1996.

88. See Madhu Kishwar, "A Himalayan Catastrophe: The Controversial Tehri Dam in the Himalayas," *Manushi,* no. 91 (November–December 1995): 5–16.

89. See Tehri Action Group, *Prime Minister Betrays Bahugana Again on Tehri Dam Review* (New Delhi: TAG, 29 August 1995).

90. See Aleta Brown, "Tehri Stalled by Powerful Protest," *World Rivers Review* 11, no. 3 (July 1996), and Amit Sengupta, "Damned If They Will," *Economic Times,* New Delhi, 24 March 1996, 7.

91. However, that the World Bank and other foreign donors chose not to become involved in the Tehri Project provides indirect evidence of the strength of the transnationally allied opposition to big dams.

92. Personal Interview with Sunderlal Bahuguna, 14 April 1996.

93. One of the earliest published sets of critiques is L. T. Sharma and Ravi Sharma, eds., *Major Dams: A Second Look* (New Delhi: Gandhi Peace Foundation-Environment Cell, 1981).

94. Ravi Sharma, "Real Cost of Big Dams" and "Need for Frank Discussion," in *Times of India,* 28 July 1982. Darryl D'Monte, "High Dams at a High Price," *Economic Scene,* Bombay, 16 November 1983. See also D'Monte, "A Question of Survival," *Illustrated Weekly of India,* New Delhi, 20–26 May 1984, 6–15.

95. Oxfam, "Dams in India," *Issue 6* (1983): 5–6; and Edward Goldsmith and Nicholas Hildyard, *The Social and Environmental Effects of Large Dams,* vol. 2 of *Case Studies* (Cornwall, U.K.: Wadebridge Ecological Centre, 1986), 201–266.

96. Society for Participatory Research in Asia, 1987.

97. See the various references to these meetings in Rich, 1994.

98. Quoted in Edward Goldsmith and Nicholas Hildyard, *The Social and Environmental Effects of Large Dams,* vol. 3 (Cornwall, U.K.: Wadebridge Ecological Centre, 1984), viii.

99. See Ramaswamy Iyer, "Large Dams: The Right Perspective," *Economic and Political Weekly,* 30 September 1989, A-107. Personal Interview with Ramaswamy Iyer, New Delhi, 23 April 1996.

100. Dhawan (1990), vii. The volume itself became controversial because it was imbalanced toward the proponents of big dams, partially because a number of opponents refused to write for the volume. An attempt to offer a balanced, but largely critical, perspective on several important aspects of the big dam debate, can be found in Dogra (1992).

101. Verghese (1994), 253.

102. Personal Interview with Shri R. Ghosh, former Chairman of the Central Water Commission 1980–82, New Delhi, 7 November 1995.

103. P. O'Neil, "India: Eternal Snows versus Finite Fuels," *International Water Power and Dam Construction* (January, 1995).

104. See "Struggle against big dams gathers momentum," *Rajasthan Patrika,* Jaipur, 11 July 1988; Sunderlal Bahugana, "Dams of Disaster: There Are Alternatives," *Hindu,* Gurgoan, 22 July 1988; and "Naturalists Force Govt. Review of Hydel Projects," *Indian Express,* Bombay, 15 August 1988.

105. Baba Amte et al., "Appeal to the Nation," Anandwan, Maharashtra, 3 July 1988.

106. Amte et al., "Assertion of Collective Will Against Big Dams," Anandwan, Maharashtra, 3 July 1988, 2.

107. See, for example, anon., "Decentralize Water Management or Perish," and "Peoples' Forests: Forest Policy, Legislation, and Practice as if People Mattered," briefs presented by nongovernmental organizations at the "Dialogue with Political Parties on Forest Lands and Water Policies," Parliament Annexe, New Delhi, 20, 22 February 1996.

108. See Ministry of Rural Development, *Rajiv Gandhi National Drinking Water Mission: A Note on Rural Water Supply Programme* (New Delhi: Government of India, 1992); Ministry of Water Resources, *National Water Policy* (New Delhi: Government of India, 1987); Narmada Bachao Andolan, *Towards a New Water Policy: Framework and Strategies* (Baroda, Gujarat: NBA, 1996); Arun Ghosh, *An Integrated Water Policy for India* (Baroda, Gujarat: NBA, 1996); H. M. Desarda, *Water Resources Development: A Note on Alternative Perspective* (Baroda, Gujarat: NBA, 1996); and Anil Agarwal and Sunita Narain, eds., *State of India's Environment 4, A Citizen's Report—Dying Wisdom: Rise, Fall and Potential of India's Traditional Water Harvesting Systems* (New Delhi: Centre for Science and Environment, 1997).

109. See National Centre for Human Settlement and Environment (NCHSE), *Documentation on Rehabilitation of Displaced Persons Due to Construction of Major Dams,* vol. 1 (Delhi: NCHSE), 18.

110. See the various pieces focusing on the draft national policy for rehabilitation in the New Delhi journal *VIKALP* 3, no. 6 (November–December, 1994). The Ministry of Environment was also pressured to reform forest policy by anti-dam and other sustainable development proponents. On the forests issue, see S. R. Hiremath et al., *All About Draft Forest Bill and Forest Lands* (Pune: Centre for Tribal Conscientization, 1995).

111. For an overview of the Harsud Rally, see Russi Engineer, "The Sardar Sarovar Controversy: Are the Critics Right?," *Business India,* 30 October–12 November 1989, 90–104. See also the discussions in on this subject in chaps. 3 and 4.

112. See the articles in *Lokayan Bulletin* 8, no. 1 (1990).

113. See National Alliance of People's Movement, *Appeal to Voters, Candidates, and the Political Parties* (Bombay: NAPM, 1994).

114. See National Alliance of People's Movements, *People's Resolve: Your Resolve* (Bombay: NAPM, 1996). The NAPM did not include all of the progressive and nonviolent social movements in India that have emerged during the 1980s and 1990s but was certainly one of the most visible in the country.

115. For analyses of India's neoliberal economic reforms, see Jeffrey Sachs et al., *India in the Era of Economic Reforms* (New Delhi: Oxford University Press, 1999).

116. See Meena Menon, "Dammed by the People: The Maheshwar Hydro-Electricity Project in Madhya Pradesh," *Business Line,* 15 June 1998.

117. See Christophe Jaffrelot, *The Hindu Nationalist Movement* (New York: Columbia University Press, 1996), and various annual editions during the 1990s of "India Review" in *Asia Survey.*

1. The Narmada is the longest river in central India, as well as the longest west-flowing and fifth longest river on the South Asian peninsula. It rises near Amarantak, in the Shahdol district of Madhya Pradesh State, at an elevation of about 2,700 feet. After traveling a distance of more than 600 miles, the Narmada River forms an approximately 22-mile-long natural border between Madhya Pradesh and Maharashtra States, after which it forms a nearly 25-mile-long natural border between Maharashtra and Gujarat States. It then flows an additional distance of about 100 miles through Gujarat and finally enters the Gulf of Khambat, draining into the Arabian Sea.

2. Geoffrey Waring Maw and Marjorie Sykes, *Narmada: The Life of a River* (Pune, India: S. J. Patwardhan, 1990); and Royina Grewal, *Sacred Virgin: Travels along the Narmada* (New Delhi: Penguin, 1994).

3. Government of India, *The First Irrigation Commission Report* (Calcutta: Government of India, 1901); and Ministry of Irrigation and Power, *Report of the Irrigation Commission* (New Delhi: Government of India, 1972), chap. 5.

4. See Ministry of Irrigation and Power, *Report of the Narmada Water Resource Committee* (New Delhi: Government of India, 1 September 1965), 1.

5. Much of the following is based on "A Historical Review of the Disputes," in *The Narmada Water Disputes Tribunal Award,* 1:15−21.

6. Quote from Ministry of Irrigation and Power (1965), 20.

7. Ministry of Works, Mines & Power, "Letter on Narmada Projects," no. 18/47 (Government of India, 19 March 1949).

8. On the importance placed on irrigation and power, see Nehru's speeches from this period in C. V. J. Varma and K. R. Saxena, *Modern Temples of India* (New Delhi: Central Board of Irrigation and Power, 1989).

9. Personal Interview with research member who expressed his wish to remain anonymous, Delhi, India, 10 October 1995.

10. Government of Bombay State, "Letter to Chairman, CWPC, on Navagam Dam," No. MIP-5559−J191249, 16 January 1959.

11. Central Waterways & Power Commission, "Letter to Bombay State on Navagam Dam," no. 7 (1)/58−FFI, 5 February 1959.

12. The creation of Gujarat and Maharashtra was part of a broader states reorganization process that took place during the 1950s and 1960s in India.

13. There is some evidence that Broach and Baroda sites were favored because politically powerful, rich farmers who were also considered to be "modern" and "progressive" among Indian authorities and bureaucrats located in these districts.

14. Government of Madhya Pradesh, *Irrigation and Power Potential of Madhya Pradesh Rivers* (Bhopal: Government of Madhya Pradesh, 1963).

15. For overviews of this period and the changing federal dynamics see Paul Brass, *The Politics of India Since Independence* (Cambridge: Cambridge University Press, 1994); and Mahendra Prasad Singh, "From Hegemony to Multi-Level Federalism?: India's Parliamentary Federal System," *Indian Journal of Social Science* 5, no. 3 (1992): 263−288.

16. There was clearly no comprehensive examination of the technical feasibility of building these three gigantic dams of 425 feet, 850 feet, and 1,390 feet at the time.

17. Government of Gujarat, "Letter to Government of India: Ratification of Bhopal Agreement," No. MIP-5563−K, 30 November 1963.

18. Government of Madhya Pradesh, Letter to Dr. K. L. Rao, Minister for Irrigation and Power from D. P. Mishra, Chief Minister, Madhya Pradesh, 28 November 1963.

19. Ministry of Irrigation and Power, "Appointment and Terms of Reference for the Narmada Water Resources Development Committee," Resolution No. DW. II-32(4)/64 (Delhi: Government of India, 5 September 1964).

20. See Ministry of Irrigation and Power (1965).

21. Francine R. Frankel, *India's Political Economy, 1947–1977: The Gradual Revolution* (Princeton: Princeton University Press, 1978); and Ashutosh Varshney, *Democracy, Development and the Countryside: Urban-Rural Struggles in India* (Cambridge: Cambridge University Press, 1995), especially chaps. 2 and 3.

22. As one Indian authority noted at the time: "In that year, Pakistan had played its usual game in Kachchh and we had to oppose them (this later on became full war in Punjab). . . . If we could provide irrigation to Kachchh, it would sustain quite a number of farmers. I, therefore, thought that we could establish our retired soldiers there. India could not financially afford a standing army in Kachchh and this was the most practical and economical way of guarding our border with Pakistan. I added my suggestion to our case which was presented to Khosla who was impressed by it." See Lalit Dalal, *All About Narmada* (Gandhinagar: Directorate of Information, Government of Gujarat, 1991), 1–2.

23. See Ministry of Irrigation and Power, 1965.

24. "Agreement Between the Governments of Madhya Pradesh and Maharashtra on the Construction of Jalsindhi Project," signed by V. P. Naik, Chief Minister of Maharashtra and D. P. Mishra, Chief Minister of Madhya Pradesh, 4 May 1965.

25. Quoted in Government of India, *Report of the Narmada Water Disputes Tribunal*, vol. 1 (New Delhi: Government of India, 1978), 20.

26. See Ministry of Irrigation and Power, *Development of Water Resources of the Narmada: Summary Record of Discussions and Conclusions Reached at the Official Level Conference* (New Delhi: Government of India, August 1966).

27. Actually, the Tribunal report mentions briefly the possible relevance of groundwater—as opposed to the flow of surface waters utilized for storage projects like dams—to an equitable apportionment of the waters of an inter-State river but then dismisses it because "groundwater flow cannot be accurately estimated from the technical point of view, and therefore, not fully cognisable as yet from the legal point of view." Government of India, vol. 1, 1978, 118. The Tribunal did cursorily examine the potential seismological impacts and rate of siltation of the Sardar Sarovar Dam.

28. In the period between the breakdown in inter-State negotiations and the constitution of the NWDT, the GOG had renamed the Navagam Project as the Sardar Sarovar Project in memory of India's nationalist hero and first Home Minister, Sardar Patel. Patel was also a Gujarati.

29. Government of Madhya Pradesh, *Before the Narmada Water Disputes Tribunal: Demurrer* (Bhopal: Government Central Press, 24 November 1969).

30. Government of Madhya Pradesh (1969), 3.

31. Volume 10 of the *Supreme Court Hearings Submissions: GOMP Response to Five Member Commission Original Report*, 15.

32. List of Relevant Dates and Events submitted by the GOG to the Supreme Court Hearing on 12 December 1995, 12.

33. Government of India (1978), 1:4.

34. Ibid., iii.

35. "Agreement Amongst the Chief Ministers of Madhya Pradesh, Gujarat, Maharashtra and Rajasthan: Narmada Development," 22 July 1972.

36. See Akhileshwar Pathak, *Contested Domains: The State, Peasantry, and Forests in Contemporary India* (New Delhi: Sage Publications, 1994), 33–40.

37. "Narmada Water Disputes Agreement," 12 July 1974. It is interesting that the conventional wisdom about Mrs. Gandhi being the cause of centralization in India's federal system during the 1970s is not borne out by this evidence. It was at the behest of the Chief Ministers that Mrs. Gandhi "intervened," and even then she returned most of the decisions to the Tribunal to adjudicate rather than "imposing" her views to solve all of the conflicts among the states.

38. Girish Patel, "Review of the Sardar Sarovar Project" (unpublished manuscript, 1994), 1–2.

39. Arch Vahini, "Displacement in SSP: A Gujarat Experience" (unpublished manuscript, Rajpipla, Gujarat, April 1988), 1–2.

40. Government of Madhya Pradesh (1969), 18.

41. Two intermediary big dams, Maheshwar and Omrakeshwar, were also planned to be built between the Narmada Sagar and Sardar Sarovar.

42. *Report of the Narmada Water Disputes Tribunal,* vol. 2 (New Delhi: Government of India, 1978), 99–107. The Tribunal did define critical terms in a rather expansive manner. "Land" included "benefits to arise out of land, and things attached to the earth." An "oustee" was anyone residing or working in any manner for at least one year in the area likely to be submerged temporarily or permanently by construction of one of the projects and "family" included the male head of family and all dependents with every major son treated as a separate family.

43. Although during this post-emergency period from 1977 until 1980 a Janata Party coalition governed in Madhya Pradesh, it was a Janata Party legislative assembly member (MLA) who first went on an indefinite hunger strike against the award, at the same time that opposition parties led by the Congress(I) strongly supported the local resistance efforts. This demonstrates the balance that politicians had to maintain between Party and State interests.

44. Anon., "Narmada award echo in M.P. Assembly," *Times of India*, 30 August 1978, 5.

45. Lathis are police batons.

46. Anon., "Protest Rally in M.P. over Narmada," *Statesman*, 8 September 1978, 1; and "Men Are Lathi Charged," *Times of India*, 8 September 1978, 1.

47. "Memorandum of Understanding," signed by Chief Minister Arjun Singh of Madhya Pradesh and Chief Minister Madhavsingh Solanki of Gujarat, 8 August 1981. The Congress (I) Party also returned to power at the federal level as well—under Mrs. Gandhi again who had brokered the 1974 agreement that had earlier unshackled the Tribunal's proceedings from the inter-State conflicts that were bogging it down at the time.

48. World Bank, *World Bank Staff Appraisal Report,* Supplementary Data Volume, pt. 1, 1985, 24–25.

49. Personal interview with Ramaswamy Iyer, 20 October 1995.

50. Tata Economic Consultancy Services, *Economic Appraisal of the Sardar Sarovar Project,* May 1983. Because the World Bank's involvement and Gujarat's interest at the time was only with the Sardar Sarovar Project, most of the studies did not focus on the Narmada Projects as a whole.

51. G.R. No. Misc. RES-1078–Amenities/Pt. 3/K-5, Gandhinagar, Gujarat, 11 June 1979.
52. Arch Vahini (1988), 3. As mentioned previously, in 1960–61, six villages had been commandeered to build Kevadia Colony, the construction headquarters for the Sardar Sarovar Dam.
53. Rajpipla Social Services Society, "A Tale of Twisted Logic: Rock Fill Dykes" (unpublished manuscript, January 1985).
54. See, for example, Chhatra Uuva Sangarsh Vahini, Rajpipla Social Service Society, and ARCH—Mangrol, "Narmada Project and the Tribals: Don't Let Them Drown in Despair," 25 February 1984.
55. The displacement issue was highlighted in the first report on *The State of India's Environment* published in 1982 by the Centre for Science and Environment, New Delhi, and by a 1984 report of Department of Tribal Development in the Government of India's Ministry of Home Affairs. Cases such as the violent resistance of indigenous peoples in the Philippines to the construction of the Chico River Dams focused international attention on the issue.
56. World Bank, "Social Issues Associated with Involuntary Resettlement in Bank-Financed Projects," Operational Manual Statement no. 2.33, February 1980.
57. World Bank, "Tribal People in Bank-Financed Projects," Operational Manual Statement, No. 2.34, February 1982.
58. Michael Cernea, "Involuntary Resettlement in Bank-Assisted Projects: A Review of the Application of Bank Policies and Procedures in FY79–85 Projects," Agriculture and Rural Development Department, World Bank, February 1986, ii.
59. Anil Patel recounts the story behind this initial letter: "So first we wrote a letter to the World Bank—a shot in the dark, but the 1982 World Bank guideline on resettlement had come out and a mission of R&R was coming in 1983 so we raised the issue of encroachers and Scudder read the letter and picked this issue up because it was a problem he had seen all over the world." Personal interview, 5 February 1996. The letter itself stated, "The oustees have been shown lands which are so poor in quality and so far away from their traditional place . . . that their economic well-being—bad as it is—would even worsen further and that their whole social and cultural life would get totally disrupted. . . . The experience of five villages which are affected by Rock Filled Dykes illustrates this point very clearly."
60. World Bank, Letter to GOI—re: need for R&R Plan, 7/1/83; Post-mission letter to GOI, GOG, GOMP & GOM—re: R&R, 7/5/83; and, Letter to GOI on need for R&R Plan, 7/25/83. The July 5th letter from William G. Rodger, Divisional Chief of Irrigation to the State Governments stated straightforwardly, "We have not received a coherent set of descriptive data . . . nor have we been provided with sufficiently detailed plans of the rehabilitation process." See also Arch Vahini, Letter to World Bank on R&R in Gujarat with respect to Sardar Sarovar Project, August 1983.
61. Bradford Morse and Thomas Berger, *Sardar Sarovar: The Report of the Independent Review* (Ottawa, Canada: Resources Futures International, 1992), 43.
62. Thayer Scudder, *The Relocation Component in Connection with the Sardar Sarovar (Narmada) Project,* November 1983.
63. See Narmada Control Authority, *Sardar Sarovar Project: Land Acquisition and Rehabilitation of Oustees* (New Delhi: Government of India, 1984).
64. Personal interview with Anil Patel of Arch-Vahini, 5 February 1996.
65. Interview with Anil Patel, 5 February 1996. Discussions with Thayer Scudder, March 1996.

66. Interview with Anil Patel. Discussion with Medha Patkar, 26 January 1996.

67. Thayer Scudder, "Aide-Memoire on the Relocation Component in Connection with the Sardar Sarovar (Narmada) Project," 21 August 1984.

68. Letter from Robin Hanbury-Tenison, President of Survival International to Ronald P. Brigish of the World Bank, 28 January 1985.

69. Letter from Thayer Scudder, Professor at California Institute of Technology to Ronald P. Brigish of the World Bank, 24 January 1985.

70. World Bank/International Development Association, *India: Narmada River Development (Gujarat) Water Delivery and Drainage Project,*" Report No. p-3938–IN, 6 February 1985, 12.

71. Letter from C. L. Robless, Chief, India Division-South Asia Department of the World Bank to Mr. Robin Hanbury-Tenison, President of Survival International on 19 April 1985. A similar correspondence between Brent Blackwelder of the Environmental Policy Institute in Washington, D.C., and senior staff at the World Bank was occurring at the same time.

72. Development Credit Agreement (Narmada River Development [Gujarat] Sardar Sarovar Dam and Power Project) between India and International Development Association, Credit Number 1552 IN, 10 May 1985.

73. Government of Gujarat—Irrigation Department, Government Resolution REHAB-Narmada-7082–48–K-5, Sachivalya, Gandhinagar, Gujarat, 30 May 1985.

74. Personal Interview with Anil Patel of Arch Vahini, 5 February 1996. Translation (out of Gujarati/Hindi) mine.

75. Government of Gujarat—Narmada Development Department, Government Resolution No. RHB-1085–D, Sachivalya, Gandhinagar, Gujarat, 1 November 1985.

76. Order of 4 February 1986 by the Supreme Court of India (Writ Petition (C) No. 7715 of 1985).

77. Government of Gujarat—Narmada Development Department, Government Resolution No. MISC-1086 (3)-D, Sachivalya, Gandhinagar, Gujarat, 21 February 1986.

78. ILO Committee of Experts, *Report of the Committee of Experts on the Application of Conventions and Recommendations* (Geneva: ILO Office, 1986), 258, 260.

79. ILO Committee of Experts, *Report of the Committee of Experts on the Application of Conventions and Recommendations* (Geneva: ILO Office, 1987), 350.

80. Ministry of Environment and Forests, Government of India, "Environmental Aspects of Narmada Sagar and Sardar Sarovar Multi-purpose Projects," April 1986.

81. Personal discussion with Medha Patkar, 26 January 1996.

82. Morse and Berger (1992), 127.

83. Ibid., 63.

84. Personal Interview with Anil Patel of Arch Vahini.

85. Letter from World Bank Director to GOI on R&R issues, 1 July 1987, and Letter from World Bank Director to GOI expressing concern about failure to provide the agreed upon 2 hectares of land with threat of suspension, 2 November 1987.

86. Government of Gujarat—Narmada Development Department, Government Resolution No. REH-1087 (66)/D, Sachivalya, Gandhinagar, Gujarat, 4 December 1987.

87. Government of Gujarat—Narmada Development Department, Government Resolution No. RHB-7087-(23)/D, Sachivalya, Gandhinagar, Gujarat, 14 December 1987.

88. Government of Gujarat—Narmada Development Department, Government Resolution No. REH—7087—CMP—12—83—D: 1 & 2, Sachivalya, Gandhinagar, Gujarat, 14 December 1987.

89. Government of Gujarat—Narmada Development Department, Government Resolution No. REH-7087-(76)/D, Sachivalya, Gandhinagar, Gujarat, 17 December 1987.

90. See Kalpavriksh, "The Narmada Valley Project: A Critique," April 1988 and Narmada Bachao Andolan, "Sardar Sarovar Project: An Economic, Environmental and Human Disaster," 1988.

4. THE TRANSNATIONAL CAMPAIGN TO SAVE INDIA'S NARMADA RIVER

1. Bradford Morse and Thomas Berger, *Sardar Sarovar: The Report of the Independent Review* (Ottawa, Canada: Resource Futures International, 1992), 220.

2. The alternatives will not be discussed in great detail in this chapter. For two moderate views, see Ashvin A. Shah, "A Technical Overview of the Flawed Sardar Sarovar Project and a Proposal for a Sustainable Alternative," in *Toward Sustainable Development: Struggling Over India's Narmada River*, ed. William F. Fisher (Armonk, N.Y.: M. E. Sharpe, 1995), 319–367, as well as Suhas Parajape and K. J. Joy, *Sustainable Technology: Making the Sardar Sarovar Project Viable* (Ahmedabad: Centre for Environment Education, 1995).

3. As Akileshwar Pathak correctly notes, "The globalisation of environmental considerations by the late sixties shifted the environmental agenda from the concerns of pollution in the developed countries to natural resources degradation in the developing countries." See his *Contested Domains: The State, Peasants, and Forests in Contemporary India* (New Delhi: Sage Publications, 1994), 32. See also "Peasants and Global Environmentalism: Safeguarding the Future of 'Our World' or Initiating a New Form of Governmentality?" in *Postcolonial Developments: Agriculture in the Making of Modern India*, by Akhil Gupta (Durham, N.C.: Duke University Press, 1998).

4. *The Fourth Five Year Plan* (New Delhi: Government of India Press, 1970).

5. Local discourses on nature and the environment have long histories in India and are not being discounted here. However, the point is that these local discourses did not produce structural changes in the Indian state; rather, transnationally constructed environmentalism did.

6. *Tiwari Committee Report on the State of India's Environment* (New Delhi: Government of India Press, 1980). The environment went on to become a central issue in India's development debate during the 1980s. See, for example, Anil Agarwal, *The State of India's Environment* (New Delhi: Centre for Science and Environment, 1982). Despite this centralizing shift in de jure authority relations with respect to environmental issues, however, the broader trend in India was still toward a decentralizing of effective power to the states. See Paul Brass, "Pluralism, Regionalism and Decentralizing Tendencies in Contemporary Indian Politics," in *The States of South Asia: Problems of National Integration*, ed. A. J. Wilson and Dennis Dalton (London: C. Hurst, 1982), 223–264, and Mahendra Prasad Singh, "From Hegemony to Multi-level Federalism? India's Parliamentary-Federal System," *Indian Journal of Social Science* 5, no. 3 (1992): 263–288.

7. The unit produced two important documents on environment and development in the mid-1970s, *Environment, Health and Human Ecologic Considerations in Economic Development Projects* and *Environment and Development*.

8. Morse and Berger (1992), 217–220. The quotes are on page 218.

9. See Paul R. Muldoon, "The International Law of Ecodevelopment: Emerging Norms for Development Assistance Agencies," *Texas International Law Journal* 22 (1986): 1–52.

10. See Kenneth Piddington, "The Role of the World Bank," in *The International Politics of the Environment,* ed. Andrew Hurrell and Benedict Kingsbury (Oxford: Clarendon Press, 1992), 217.

11. Kalpavriksh—The Environmental Action Group and the Hindu College Nature Club, Delhi University; "The Narmada Valley Project—Development or Destruction?" (unpublished manuscript, September 1983), 8.

12. Kalpavriksh (1983), 30–31.

13. Kothari conveyed that Goldsmith was a friend of the family and thus he felt comfortable writing him about the work Kalpavriksh had conducted on the Narmada Projects. Interview with Ashish Kothari, New Delhi, 10 December 1995.

14. Edward Goldsmith and Nicholas Hildyard, eds., The *Social and Environmental Effects of Large Dams* (Cornwall, U.K.: Wadebridge Ecological Center, 1986).

15. Interview with Dr. S. Maudgal, Senior Advisor to the Ministry of Environment and Forests, New Delhi, 14 November 1995.

16. See Sanat Mehta, *Further Delay in Narmada Project (SSP) Will Provide Disastrous For Gujarat: (A detailed case for early clearance)* (Baroda: Gujarat Foundation for Development Alternatives, 1986).

17. Interview with T. N. Seshan, New Delhi, 10 December 1995.

18. Medha Patkar (in conversation with Smitu Kothari), "The Struggle for Participation and Justice: A Historical Narrative," in *Toward Sustainable Development: Struggling Over India's Narmada River,* ed. William F. Fisher (New York: M. E. Sharpe, 1995), 157–160. T. N. Seshan was the Secretary of the MOEF from 1985 to 1987.

19. Interview with Medha Patkar, Narmada River Valley in Madhya Pradesh, 26 January 1996. Himanshu Thakkar, a leading activist of the Narmada Bachao Andolan, suggested that environmental issues came to the forefront as a result of four factors: (1) the environmental clearance process through the GOI's Ministry of Environment and Forests; (2) the Hindu Nature Club/Kalpavriksh report written by Ashish Kothari; (3) the growing strength of environmental NGO's and the environmental movement in India; and (4) the involvement of transnational environmental NGO's like the Environmental Defense Fund and International Rivers Network. Interview with Himanshu Thakkar, New Delhi, 22 January 1996.

20. Lori Udall, "The International Narmada Campaign: A Case Study of Sustained Advocacy," in *Toward Sustainable Development,* 202–203.

21. Udall, in Fisher (1995), 202–203.

22. Interview with Medha Patkar, Narmada River Valley in Madhya Pradesh, 26 January 1996.

23. Key members of the Action Committee included Medha Patkar (Narmada Bachao Andolan); Yukio Tanaka (Friends of Earth, Japan); Carol Sherman (Rainforest Information Center, Australia); Bruni Weisen (Action for World Solidarity); Paul Wolvekamp (Both Ends); Risto Isoamki (Finland); Ville Komsi (Committee on Environment and Development); Frank Brassel (FIAN, Germany); Nicholas Hildyard and Alex Wilkes (The Ecologist, England); Patrick McCully (formerly at the Ecologist, England, then with International Rivers Network, California); Aditi Sharma (Survival International, England); Christian Ferrie (France); Peggy Hallward and Pat Adams (Probe International, Canada); Maud Joahanssen and Joran Ecklaf (The Swallows, Sweden); Juliet Majot (International Rivers Network, California); and Lori Udall (formerly of the Environmental Defense Fund, Washington, D.C., and later with International Rivers Network, Washington, D.C.).

24. Patkar in conversation with Smitu Kothari in Fisher (1995), 157–160.

25. Interview with Dr. S. Maudgal, Senior Advisor to the Ministry of Environment and Forests, New Delhi, 14 November 1995.

26. Ibid.

27. Ibid. Interview with Ramaswamy Iyer, the Secretary of the Ministry of Water Resources from 1984 to 1986, New Delhi, 23 April 1996.

28. Interview with Chief Minister of Gujarat Chimanbhai Patel, Gandhinagar, 2 April 1993. By "lives" Patel seemed to be referring both to the political lives of elected officials as well as the "millions of Gujaratis" he thought would benefit from the project (translated from Gujarati.)

29. Interview with Ramaswamy Iyer, the Secretary of the Ministry of Water Resources from 1984 to 1986, New Delhi, 23 April 1996.

30. Ministry of Environment and Forests, Environmental Clearance for Narmada Sagar and Sardar Sarovar Projects, Office Memorandum no. 3–87/80–1A, Government of India, New Delhi, 24 June 1987.

31. See Kalpavriksh, "The Narmada Valley Project: A Critique," April 1988 and Narmada Bachao Andolan, "Sardar Sarovar Project: An Economic, Environmental and Human Disaster," 1988.

32. This is not to support the view in any way that the organizing efforts in Gujarat and Maharastra were easy, as Amita Baviskar has inferred in her analysis *In the Belly of the River: Tribal Conflicts Over Development in the Narmada Valley* (New Delhi: Oxford University Press, 1995).

33. Interview with various tribal and nontribal villagers, Narmada River Valley in Madhya Pradesh, March 1993.

34. A major critique of the projects that had been written by two independent researchers that gave a huge boost to the domestic struggle and heightened international attention was published in May 1988. See Claude Alvares and Ramesh Billorey, *Damming the Narmada: India's Greatest Planned Environmental Disaster* (Penang, Malaysia: Third World Network, 1988).

35. Arch Vahini, "The 14th May, 1988 Vadgam Convention—Before and After," 21 March 1988, 9.

36. Patkar in conversation with Smitu Kothari, in Fisher ed. (1995), 161–162.

37. Arch Vahini, 1988.

38. For further analysis of the split, see Joelle Tamraz, "Gandhi and the Narmada Controversy: A Discussion on the Means and Ends of Economic Development," An Essay Presented to The Committee on Degrees in Social Studies, Radcliffe College, March 1995 and Navroz K. Dubash, "The Birth of an Environmental Movement: The Narmada Project as Seed-Bed for Civil Society in India" (a Senior Thesis Presented to the Faculty of the Woodrow Wilson School of Public and International Affairs, Princeton University, 9 April 1990).

39. Rs refers to Indian rupees, and crore is the Indian term equivalent to 10 million.

40. See Girish Patel, "Sardar Sarovar Project and the Official Secrets Act" (unpublished manuscript, n.d.).

41. Interview with Sitarambhai, New Delhi, 21 January 1996 (translation from Hindi). Himanshu Thakkar, a leading activist, further stated that, "The governments tried—they imposed the Official Secrets Act for the first time in a river valley, that was never done before. They harassed people in the villages, they arrested activists, they tried to stop pro-

grams from happening, they committed human rights abuses, they tried everything to stop the movements. They could not." Interview, New Delhi, 21 January 1996.

42. Vijay Paranjpye, *High Dams on the Narmada: A holistic analysis of River Valley Projects* (New Delhi: Indian National Trust for Art and Cultural Heritage, 1990), 29–30.

43. Interview with Sripad Dharmadikary, Baroda-Gujarat, 3 February 1996.

44. See Smitu Kothari, "Special Issue on Dams on the River Narmada," *Lokayan Bulletin* 9, nos. 3/4 (1991). An alternative and more participatory democratic process was also taking shape in the Narmada Valley along side the fight against destructive development. For example a Karbhar Samiti in which two representatives from every village, one of which had to be a woman, met during the full moon of every month to debate and decide on issues facing them was established.

45. Interview with Himanshu Thakkar, New Delhi, 22 January 1996.

46. Moeen Qureshi, Letter to Arjun Singh, Chief Minister of Madhya Pradesh, 28 November 1988.

47. Thayer Scudder, Memorandum to Michael Baxter of the World Bank, 29 April 1989.

48. Udall, in Fisher (1995), 210.

49. United States House of Representatives Committee on Science, Space, and Technology, *Sardar Sarovar Dam Project: Hearing Before the Subcommittee on Natural Resources, Agricultural Research and Environment,* 101st Congress, 1st sess. 24 October 1989 (Washington, D.C.: U.S. Government Printing Office, 1990).

50. Udall, 211.

51. United States House of Representatives Committee on Science, Space, and Technology (1989, 1990), 206.

52. Udall, 206.

53. See the series of e-mail and fax correspondence between Friends for Earth—Japan and International Rivers Network—Berkeley, California, 1989–1991.

54. Udall, 212.

55. Letter from Heinz Virgin, Director, India Country Department to M. S. Reddy, Secretary, Ministry of Water Resources, Government of India, February 1990.

56. Interview with Baba Amte, 12 May 1994.

57. Interview with former Chief Minister of Gujarat Chimanbhai Patel, Gandhinagar, 2 April 1993 (translation from Gujarati).

58. See "Before the Deluge: Human Rights Abuses at India's Narmada Dam," *Human Rights Watch/Asia* 4, issue 15 (17 June 1992).

59. Baviskar (1995), 207.

60. Interview with Manuben, Rajpipla, 10 April 1993 (translation from Gujarati.)

61. Interview with Sripad Dharmadikary, Baroda-Gujarat, 3 February 1996.

62. Interviews with Lori Udall and Patrick McCully of the International Rivers Network, Washington, D.C., and Berkeley, California, April 1994 and March 1996.

63. A precedent was in fact set and a World Bank Inspection Panel was subsequently established in 1993. As Ibrahim F. I. Shihata later wrote, "The single most important case to draw public attention to the accountability issue involved two major projects supported by the World Bank on the Narmada River in India. . . ." *The World Bank Inspection Panel* (Oxford: Oxford University Press, 1994), 9–10.

64. Interview with Medha Patkar, India, February 1996.

65. Patkar in conversation with Smitu Kothari, in Fisher (1995), 171.

66. Personal interviews in the Narmada River Valley, April 1993.

67. See Narmada Control Authority, "Minutes of the Meetings of the Environmental Sub-Group," Meetings 10 to 19 (Delhi: Narmada Control Authority, February 1991 to July 1993).

68. Other reports on human rights abuses included for example: Chris A. Wold, *Narmada International Human Rights Panel—Interim Report,* 7 October 1992; Human Rights Campaign on Narmada, *Respect Human Rights: Violations of Human Rights in the Narmada Valley,* 10 October 1993; and Lawyers Committee for Human Rights, *Unacceptable Means: India's Sardar Sarovar Project and Violations of Human Rights,* New York: Lawyers Committee For Human Rights, April 1993.

69. As one villager in the Narmada Valley who asked to remain anonymous later told me, "Yes the big men from the United Nations came and wanted to see how we live in our villages. The waters of the river were running high so it was difficult for them to cross. So we offered that we could carry them across on our backs. They said they could not accept such an offer. We told them, 'don't worry so much, we always have carried the World Bank on our backs.'" Personal Interview, Narmada River Valley, May 1993.

70. Morse and Berger (1992), 357–358.

71. Ibid., 349–356.

72. *Comment on the Report of the Independent Review Mission on Sardar Sarovar Project* (Gandhinagar: Government of Gujarat, August 1992).

73. Narmada Foundation Trust and Gujarat Chamber of Commerce and Industry, *An Open Letter to Mr. Lewis T. Preston, President, The World Bank, on Continuing Aid for the Sardar Sarovar Project* (Ahmedabad, Gujarat: Gujarat Chamber of Commerce and Industry, 11 August 1992).

74. See Pravin Seth, *Narmada Project: Politics of Eco-Development* (New Delhi: Har-Anand Publications, 1994), especially 80–86.

75. Letter from M. A. Chitale, Secretary, Government of India Ministry of Water Resources, "Independent Review," to Mr. Heinz Virgin, Chief, Indian Operations Department, World Bank, 7 August 1992.

76. Interview with Sripad Dharmadikary, Baroda-Gujarat, 3 February 1996.

77. Narmada International Action Committee, "The World Bank Must Withdraw Immediately from Sardar Sarovar: An Open Letter to Mr. Lewis T. Preston, President of the World Bank," *Financial Times,* 21 September 1992, 5.

78. Bradford Morse and Thomas Berger, Letter to Mr. Lewis T. Preston, sent by fax, 13 October 1992.

79. See World Bank, "India: Sardar Sarovar (Narmada) Projects: Management Response," SecM92–849, 23 June 1992. See also T. R. Berger, "The World Bank's Independent Review of India's Sardar Sarovar Projects," *American University Journal of International Law and Policy* 33 (1993).

80. Personal interviews and observations, Narmada River Valley, April–June 1993.

81. Quoted in Anumita Roychowdhury and Nitya Jacob, "Was India forced to reject World Bank aid?" *Down to Earth,* 30 April 1993, 5.

82. Quoted in ibid., 7.

83. Many of these NRI's had become wealthy after immigrating to their new countries yet still had strong social and cultural ties to Gujarat and India. See various issues of *India Abroad* and *India West,* newspapers published for the NRI community living outside of India from 1990 on.

84. See Narmada Control Authority, "Minutes of the 20th Meeting of the Environmental Sub-Group," Indore, 7 December 1993.

85. Narmada Bachao Andolan, "Press Release: Samarpit Dal of Narmada Valley to Embrace Narmada Water in Manibeli Against Submergence" and "Narmada Samachar—Narmada News," Baroda: Narmada Bachao Andolan, 13 April and September 1992.

86. See Nitya Jacob, "Narmada Review: Break or Breakthrough?" *Down to Earth*, 15 July 1993, 5–8.

87. Letter from N. Suryanarayanan, Commissioner of Policy and Planning, Ministry of Water Resources to Narmada Bachao Andolan, 17 July 1993; Letter from Sripad Dharmadhikary, Narmada Bachao Andolan to V. C. Shukla, Minister of Water Resources, 23 July 1993; Letter from Shri Ahuja, Commissioner, Ministry of Water Resources to Medha Patkar, Narmada Bachao Andolan, 27 July 1993; Letter from Sripad Dharmadhikary, Narmada Bachao Andolan to N. Suryanarayanan, Commissioner of Policy and Planning, Ministry of Water Resources, 30 July 1993.

88. Government of India, Ministry of Water Resources, Office Memorandum: Constitution of a Group to Continue Discussions on Sardar Sarovar Project, 3 August 1993.

89. *The Report of the Five-Member Group* (New Delhi: Government of India, April 1994), 1–2, and Government of India, Ministry of Water Resources, Office Memorandum: Constitution of a Group to Continue Discussions on Sardar Sarovar Project, 5 August 1993.

90. *The Report of the Five-Member Group* (1994), 3–5.

91. "Sardar Sarovar Dam to Be Taken to 80 Metres," *Times of India*, New Delhi, 19 January 1993.

92. See Narmada Control Authority (1993) and *The Report of the Five-Member Group* (1994), 9. Shortly thereafter, the MOEF released 1500 hectares of reserved forest lands for resettlement, once again in direct violation of legal procedure.

93. Interview with tribal leader, Barwani, Madhya Pradesh, 14 February 1996 (translated from Hindi).

94. *In the Supreme Court of India: Writ Petition (Civil) No. 319 of 1994, Narmada Bachao Andolan versus The Union of India* (1994), 1–2.

95. Supreme Court of India, *Order of The Supreme Court of India on Writ Petition N.319/94*, 24 January 1995.

96. See, *Further Report of the Five-Member Group*, 2 vols. (New Delhi: Government of India, 16 April 1995), especially 1: 109–124.

97. This section is based on personal attendance at the Supreme Court Hearings: Narmada Bachao Andolan versus Union of India, New Delhi, November 1995, January and April 1996, and February 1997 as well as written submissions of various parties to the case.

98. During the late 1990s the movement in the valley was given a big boost when award-winning writer Arundhati Roy began to actively support them. However, while Roy was clearly important in boosting the morale of domestic opponents and once again raising the visibility of the anti-Narmada campaign around the world, her involvement seemed also to provoke even greater organization, mobilization, and violence by pro-dam actors, particularly extreme Hindu fundamentalist groups and leaders.

99. In contrast, the lack of transnationally allied lobbying and strong domestic mobilization resulted in the completion of the Bargi dam in the Madhya Pradesh part of the Narmada Valley in 1990. That project cost many times more than was projected, and those displaced were not given even minimally adequate resettlement and rehabilitation packages. Those displaced from the Bargi dam thus continued to struggle for reparations and contributed to the broader movements opposing big dams in the Narmada Valley, across India and around the world throughout the 1990s.

1. See Larry Diamond, "The Globalization of Democracy," in *Global Transformation and The Third World*, ed. Robert O. Slater et al. (Boulder: Lynn Rienner, 1993); and Philippe C. Schmitter, "Democracies' Future: More Liberal, Preliberal, or Postliberal?" *Journal of Democracy* 6, no. 1 (1995).

2. The comparison is also useful because it involves countries in which the decreasing number of available sites is clearly part of the overall decrease in big dam building. Nevertheless, these countries continue to plan and fund these projects; as a result, the transnational political-economic explanation proposed in this study for the decreasing ability to implement them is strongly corroborated.

3. See ICOLD, *World Register of Dams* (Paris: ICOLD, 1998), 57.

4. See Shelton Davis, *Victims of Miracle* (Cambridge: Harvard University Press, 1977), 177.

5. See Bolivar Lamounier, "Brazil: Inequality Against Democracy," in *Politics in Developing Countries: Comparing Experiences with Democracy*, ed. Larry Diamond et al. (Boulder, Colo.: Lynne Rienner, 1995), 140.

6. See Catherine Caufield, *Masters of Illusion: The World Bank and the Poverty of Nations*, 84.

7. See Thomas E. Skidmore, "Politics and Economic Policy Making in Authoritarian Brazil," in *Authoritarian Brazil*, ed. Alfred Stepan (New Haven: Yale University Press).

8. See Mark D. McDonald, "Dams, Displacement, and Development: A Resistance Movement in Southern Brazil," in *In Defense of Livelihood: Comparative Studies on Environmental Action*, ed. John Friedman and Haripriya Rangan (West Hartford, Conn.: Kumarian Press, 1993), 82.

9. See Bruce Rich, *Mortgaging the Earth: The World Bank, Environmental Impoverishment, and the Crisis of Development* (Boston: Beacon, 1996), 99–100.

10. See Stephen Schwartzman and Michelle Malone, "MDBs and the Brazilian Electrical Energy Sector," in *Hydroelectric Dams on Brazil's Xingu River and Indigenous Peoples*, ed. Leinad Ayer de O. Santos and Lucia M. M. de Andrade (Cambridge, Mass.: Cultural Survival, 1990), 62–65.

11. See Robert Goodland, "Brazil's Environmental Progress in Amazonian Development," in *Man's Impact on Forests and Rivers*, ed. J. Hemmings (Manchester: Manchester University Press, 1983), 6, and "Environmental Ranking of Amazonian Development Projects in Brazil," *Environmental Conservation* 7, no. 1 (Spring 1980): 9–10.

12. Lucia Andrade and Leinad Ayer Oliveira Santos, *Project Dossier: Hydroelectrics of the Xingu and Indigenous People* (São Paulo, Brazil: Pro-Indian Commission of São Paulo, 1988).

13. See Patrick McCully, *Silenced Rivers: The Ecology and Politics of Large Dams* (London: Zed Books, 1996), 70, 243. See also Philip M. Fearnside, "Brazil's Balbina Dam: Environment versus the Legacy of the Pharaohs in Amazonia," *Environmental Management* 13, no. 4: 401–423, as well as Eduardo Viveiros de Castro and Lucia M. M. de Andrade, "Xingu Hydroelectrics: The State Versus Indigenous Societies," in Santos and Andrade (1990), 1–16.

14. See Aurelio Vianna and Federico Guilherme Bandeira de Araujo, eds., *Terra Sim, Barragens Não: First Encontro Nacional de Trabalhadores Atingidos por Barragens* (Brasil: Central Unica dos Trabalhadores e Comissão Regional de Atingidos por Barragens, October 1989), 16–19.

15. See Ricardo Canese and Luis Alberto Mauro, *Itaipu: Dependencia o Desarrollo* (Asuncion del Paraguay: Azara, 1985).

16. See McCully (1996), 21, 271, and 86.

17. See Vianna and Araujo (1989), 25.

18. See McCully (1996), 294.

19. See Lamounier in Diamond et al., 139.

20. See Peter P. Houtzager, *Caught between State and Church: Popular Movements in the Brazilian Countryside, 1964–1989* (Ph.D. diss., Department of Political Science, University of California, Berkeley, 1998) for an analysis of the rise of popular movements during the authoritarian and transition periods in Brazil.

21. Stephen P. Mumme and Edward Korzetz, "Democratization, Politics, and Environmental Reform," in *Latin American Environmental Policy in International Perspective,* ed. Gordon J. MacDonald, Daniel L. Nielson, and Marc A. Stern (Boulder, Colo.: Westview, 1997), 43–44.

22. See Raul Pont, *Da crítica ao populismo à construção do PT* (Porto Alegre: Ed. Seriema, 1985); Candido Grzybowski, *Caminhos e descaminhos dos movimentos sociais no campo* (Petrópolis: Vozes, 1987); Ilse Scherer-Warren e Paulo Krischke, *Uma revolução no cotidiano? Os novos movimentos sociais na América Latina* (São Paulo: Brasiliense, 1987); Vito Giannotti e Sebastião Neto, *CUT, por dentro e por fora* (Petrópolis: Vozes, 1990), Emir Sader, *A transição no Brasil: da ditadura à democracia?* (São Paulo: Ed. Atual, 1990); and Bernardo Mançano Fernandes, *A formação do MST no Brasil* (Petrópolis: Vozes, 2000).

23. See also Allison Brysk, "Turning Weakness into Strength: The Internationalization of Indian Rights," *Latin American Perspectives,* 23, no. 2 (1996): 38–57.

24. See Brazilian Embassy, *Brazilian Policy on Indigenous Population* (Washington, D.C.: Brazilian Embassy in Washington Documentation Center, 1993).

25. See Goodland in Hemmings (1983), 1.

26. Eduardo Viola, "The Environmental Movement in Brazil: Institutionalization, Sustainable Development, and Crisis of Governance Since 1987," in MacDonald, Nielson, and Stern (1997), 94–97.

27. See Vianna and Araujo (1989), 25–26.

28. Ibid., 27.

29. The following account on CRAB and the Uruguai Basin Projects is partially based on Comissão Regional de Antigidos por Barragens, *Nossa historia em debate, Mais de dez anos de luta Erexim* (Rio Grande do Sul, Brazil: CRAB, 1989); McDonald in Friedman and Rangan (1993), 79–105; and a series of interviews conducted with members of CRAB in Curitiba and São Paulo, Brazil, in February and March 1997.

30. See Rich (1994), 136–137.

31. See Celio Berman, "Self-Managed Resettlement—A Case Study: The Ita Dam in Southern Brazil" (paper presented at the Conference on Hydropower Into the Next Century, Barcelona, Spain, June 1995); and A. Oliver Smith, "Fight for a Place: The Policy Implications of Resistance to Development-Induced Resettlement" (paper presented at the Conference on Development-Induced Displacement and Impoverishment," Oxford, England, January 1995).

32. McDonald in Friedman and Rangan (1993), 99–100.

33. See Ana Luiza B. Martins Costa et al., *Hidroeletricas, ecologia e progresso* (Rio de Janeiro, Brazil: Centro Ecumenico de Documentacao e Informacao, 1990), especially 55–59, and Vianna and Araujo (1989), 16.

34. Interview with Fulgencio Silva, leader of the peoples affected by the Itaparica Dam, Curitiba, Brazil, 14 March 1997. See also Costa et al. (1990), especially 55–59; and Vianna and Araujo (1989), 16–19.

35. See Schwartzman and Malone in Santos and Andrade (1990), 62–65; Vianna and Araujo (1989), 16–19; and Rich (1994), 136–137.

36. See Goodland in Hemmings (1983), 6, and "Environmental Ranking of Amazonian Development Projects in Brazil," *Environmental Conservation* 7, no. 1 (Spring 1980): 9–10. See also Elizabeth Monosowski, "Brazil's Tucurui Dam: Development at Environmental Cost," in *The Social and Environmental Effects of Large Dams*, vol. 2, ed. Edward Goldsmith and Nicholas Hildyard (Camelford, U.K.: Wadebridge Ecological Centre, Case Studies, 1986), 191–198.

37. Vianna and Araujo (1989), 10–13.

38. See Castro and Andrade in Santos and Andrade (1990), 8–9.

39. Personal interviews with dam affected in Belem, Brazil, March and April 1997.

40. Vianna and Araujo, 13.

41. See Fearnside, 417.

42. The following discussion is heavily indebted to the following sources: Vianna and Araujo, 13–15; it is also based on interviews conducted by the author. Castro and Andrade in Santos and Andrade (1990); and McCully (1996), 292–294.

43. See Andrade and Santos (1988). Interview with Owen Lammers, International Rivers Network, Berkeley, California, 5 June 1995.

44. See McCully (1996), 293.

45. Interview with Ricardo Montaigner, MAB, Curitiba, Brazil, 15 March 1997. See Ricardo Azambuja Arnt and Stephan Schwartzman, *Um Artificio Organico: Transicao Na Amazonia E Ambientalismo* (Rio de Janeiro: Rocco, 1992), 356–360; Vianna and Araujo 13–15; and Nicholas Hildyard, "Adios Amazonia?: A Report from the Altimira Gathering," *Ecologist* 19, no. 2 (1989).

46. *Quilombos* are communities of the descendants of Africans who escaped slavery in Brazil between the sixteenth and nineteenth centuries.

47. See Luiz Alencar Dalla Costa, "Barragens: da Politica Oficial a Resistencia dos Atingidos," in *Travessia* (Rio de Janeiro, Brazil, April 1990), 3, 47–51; Vianna and Araujo (1989), 37.

48. Anon., *Atingidos Constroem Uma Luta Unificada A Nivel Nacional e Nasce O MAB (Movimento Dos Atingidos Por Barragens* (São Paulo: MAB, 1995).

49. Eletrobrás, *Plano Nacional de Energia Eletrica—1987/2010—Plano 2010* (Rio de Janeiro, Relatorio Geral, 1987). This was confirmed during interviews conducted by the author with various senior officials in the federal government in August 1999.

50. Anon., "Brazil's Dam Plan Falls Behind Schedule," *World Rivers Review* 6, no. 5 (July/August 1991): 5, 9; and Carlos Vanier, "The Past and Continuing Struggle Against Dams" (paper prepared for the First International Conference of Peoples Affected by Dams, Curitiba Brazil, 11–14 March 1997).

51. Colin MacAndrews, "Politics of the Environment in Indonesia," *Asian Survey* 34, no. 4 (April 1994): 369–370.

52. See George Aditjondro and David Kowalewski, "Damning the Dams in Indonesia," *Asian Survey* 34, no. 4 (April 1994): 384.

53. See Rich (1994), 100.

54. See MacAndrews (1994), 374.

55. See James V. Riker, "Linking Development from Below to the International Environmental Movement: Sustainable Development and State-NGO Relations in Indonesia," *Journal of Business Administration* 22–23 (1994/1995): 162–163.

56. For an overview, see Philip J. Eldridge, *Non-Government Organizations and Democratic Participation in Indonesia* (Oxford: Oxford University Press, 1995).

57. See James V. Riker, "NGOS, Transnational Networks, International Donor Agencies and the Prospects for Democratic Governance in Indonesia," in Khagram, Riker, Sikkink, *Restructuring World Politics: Transnational Social Movements, Networks, and Norms* (Minneapolis: University of Minnesota Press, 2002), 181–205; and Adnan Buyung Nasution, "Defending Human Rights in Indonesia," *Journal of Democracy* 5, no. 3 (1994): 116.

58. See Riker (1994/95), 165; and Riker (1996), 3–7. The consortium of foreign donors is the Inter-Governmental Group on Indonesia (IGGI) and INFID has changed names several times; it was originally the Inter-NGO Conference on IGGI Matters (INGI).

59. See Aditjondro and Kowalewski (1994), 384–391.

60. Ibid., 385–391.

61. Ibid., 394.

62. See Seamus Cleary, *The Role of NGOs under Authoritarian Political Systems* (New York: St. Martin's, 1997), 14–25.

63. See MacAndrews (1994), 372, 370.

64. Anon., *In the Name of Development: Human Rights and the World Bank in Indonesia: A Joint Report of The Lawyer's Committee for Human Rights and The Institute for Policy Research and Advocacy (ESLAM)* (New York: Lawyer's Committee for Human Rights, July 1995), 93.

65. This analysis of the Kedung Ombo cases is heavily indebted to the following sources: Ibid.; Augustinnus Rumansara, "Indonesia: The Struggle of the People of the Kedung Ombo," in *The Struggle for Accountability: The World Bank, NGOs, and Grassroots Movements*, ed. Jonathan Fox and L. David Brown (Cambridge: MIT Press, 1999); Rich (1994), 149–150; Cleary (1997), 25–37; George Aditjondro, "The Impact of the Kedung Ombo Dam Campaign" (unpublished manuscript, 1992); and World Bank, *Project Completion Report: Indonesia, Kedung Ombo Multipurpose Dam and Irrigation Project* (Washington, D.C.: World Bank, October 1994).

66. See ICOLD (1998), 143.

67. See International Hydropower Association, *Hydropower and Dams World Atlas* (1997), 45.

68. See Riker (2002).

69. See ICOLD (1998), 243.

70. The following is partly based on World Commission on Dams, *Orange River Development Project, South Africa* (World Commission on Dams Case Study: Cape Town, 1999), and personal interviews conducted in Lesotho and South Africa in 1998–99.

71. For an overview of Lesotho's military regime between 1986 and 1990, see Stephen J. Gill, *A Short History of Lesotho* (Morija, Lesotho: Morija Museum and Archives, 1993), 240–246. For an overview of South Africa's Apartheid regime, see William Beinart, *Twentieth-Century South Africa* (Oxford: Oxford University Press, 1994).

72. See Anon., *Lesotho Highlands Water Project* (Berkeley, Calif.: International Rivers Network, June 1996); and Anon., *Power Conflicts: Norwegian Hydropower Developers in the Third World* (Oslo, Norway: FIVAS [Association for International Forest and Water Studies], 1996), chap. 11.

73. *Treaty on the Highlands Water Project,* Article 10, Paragraph 3h.

74. See Mathato Khits'ane, "Highlands Church Action Group in the Affected Areas by the Lesotho Highland Water Project" (paper presented at the First International Conference of Dam-Affected Peoples, Curitiba, Brazil, 11–14 March, 1997).

75. See Roger Southall, "Lesotho's Transition and the 1993 Election," in *Democratisation and Demilitarisation in Lesotho: The General Election of 1993 and Its Aftermath,* ed. Roger Southall and Tsoeu Petlane (Pretoria: Africa Institute of South Africa, 1995), chap. 2.

76. Interview with Mrs. Mathato Khits'ane of HCAG, Curitiba, Brazil, 12 March 1997.

77. See Robert Archer, *The Lesotho Highlands Water Project* (London: Christian Aid, January 1994).

78. Quoted in *Lesotho Highlands Water Project* (January, 1994).

79. See letter to Mr. Edward V. K. Jaycox, Vice President, Africa Region Office, The World Bank from Christa Coleman of International Rivers Network and Korinna Horta of the Environmental Defense Fund, 28 September 1995, and Robert Archer, *Trust in Construction?: The Lesotho Highlands Water Project* (London: Christian Aid, March 1996).

80. See Lori Pottinger, "Lesotho Water Project Awash in Troubles," in *World Rivers Review* (Berkeley, Calif.: International Rivers Network, April 1996), 4; letter to Mr. John Roome, Lead Specialist, Infrastructure, Africa Region, World Bank from Mzwanele Mayekiso, Co-Chair, Ad hoc Committee, SANCO Greater Johannesburg Region, 26 November 1997; Abid Aslam, *Romawas Development: South Africans Clash with World Bank Over Water* (Washington, D.C.: Interpress Service, 10 March 1998); letter to Jean-Louis Sarbib, Vice President, Africa Department, The World Bank from Ms. Korinna Horta, Environmental Defense Fund and Ms. Lori Pottinger, International Rivers Network, 7 March 1997; and letter responding to Ms. Korinna Horta, Environmental Defense Fund and Ms. Lori Pottinger, International Rivers Network from Callisto E. Madavo, Vice President, Africa Region, The World Bank, 10 April 1997.

81. This was the case until a major corruption scandal and corresponding court case surrounding the construction of the LWHP emerged in August 1999. See George Dor, "The privatisation of utilities is an invitation to bribery and graft," *Business Day,* August 17, 1999.

82. See Hein Marais, *South Africa: Limits to Change, The Political Economy of Transition* (London: Zed Books, 1999).

83. Department of Water Affairs and Forestry, *White Paper on Water Policy* (Pretoria: Government of the Republic of South Africa, 1997), 28. The National Water Act, passed in 1998, included similar language.

84. Shui Fu, "A Profile of Dams in China," in *The River Dragon Has Come!: The Three Gorges Dam and the Fate of China's Yangtze River and Its People,* ed. Dai Qing (London: M. E. Sharpe, 1998), 18–22.

85. See Michael Oksenberg, *Policy Formulation in China: The Case of the 1957–58 Water Conservancy Campaign* (Ph.D. diss., Columbia University, 1969).

86. For an overview of the Maoist strategy of economic development, see John G. Gurley, *China's Economy and the Maoist Strategy* (New York: Monthly Review Press, 1976).

87. See Fu in Qing (1998), 22–23; Yi Si, "The World's Most Catastrophic Dam Failures," in Qing (1998), 25–38, and McCully (1996), 8, 66–67.

88. See Anon., "Hydropower & Dams World Atlas: China," *International Journal of Hydropower and Dams* (1997).

89. For an overview, see He Bochuan, *China on the Edge: The Crisis of Ecology and Development* (San Francisco: China Books and Periodicals, 1991); and Vaclav Smil, *China's*

Environmental Crisis: An Inquiry into the Limits of National Development (Armonk, N.Y.: M. E. Sharpe, 1993).

90. See Kenneth Lieberthal and Michael Oksenberg, *Policy Making in China: Leaders, Structures, and Processes* (Princeton: Princeton University Press, 1988); David Lampton and Kenneth Lieberthal, eds., *Bureaucracy, Politics, and Decision Making in Post-Mao China* (Berkeley: University of California Press, 1992); and A. L. Brown, R. A. Hindmesh, and G. T. McDonald, "Environmental Assessment Procedures an Issues in the Pacific Basin—Southeast Asia Region," *Environmental Impact Assessment Review* 11: 143–156.

91. See Susan Whiting, *The Non-governmental Sector in China: A Preliminary Report* (Beijing, China: Ford Foundation, July 1989); and Anna Brettell, "Environmental Non-governmental Organizations in the People's Republic of China: A Comparative Perspective" (paper presented at International Studies Association Conference, Minneapolis, Minnesota, 18–21 March 1998).

92. See Barrett McCormick, "The 1989 Democracy Movement: A Review of the Prospects for Civil Society in China," *Pacific Affairs* 63, no. 2 (Summer 1992); and Jeffrey Wasserstrom and Elizabeth Perry, *Popular Protest and Political Culture in Modern China* (Boulder, Colo.: Westview, 1994).

93. See Dai Qing, "The Three Gorges Project: A Symbol of Uncontrolled Development in the Late Twentieth Century," in Qing, 3.

94. Much of the following is indebted to Grainne Ryder, "Damming the Three Gorges: 1920–1990," in *Damming the Three Gorges: What Dam-Builders Don't Want You to Know* (Toronto: Probe International, 1991), 16–28; and Tian Fang and Lin Fatang, *A Third Look at a Long-Range Strategy for the Three Gorges Project* (Hunan: People's Publishing House of Hunan, 1992).

95. Poem by Mao Zedong, translated by John Gittings, provided by Ramachandra Guha.

96. For an analysis of these conflicts, see Lieberthal and Oksenberg (1988).

97. See Dai Qing, "The Struggle to Publish the *Yangtze! Yangtze!* in China," in *Yangtze! Yangtze!*" ed. Patricia Adams and John Thibodeau (London: EarthScan Publications, 1994), 2.

98. Quoted in Patricia Adams and Philip Williams, "Introduction: Opposition to an Unviable Dam," in *Yangtze! Yangtze!*, xxiii.

99. Quoted in Anon., "Vigorous Debate Delays China Dam," *Christian Science Monitor*, 1 March 1989, 6. The following is based on Shi He and Ji Si, "The Comeback of the Three Gorges Dam," in Qing, 22–45. Interview with Dai Qing, Santa Cruz, California, 14 November 1997.

100. Interviews with Owen Lammers, International Rivers Network, Berkeley, California, 7 January 1998 and Dai Qing, Santa Cruz, California, 14 November 1997. See also Qing (1998).

101. Audrey Ronning Topping, "Cracking the Wall of Silence," *New York Times Magazine*, 5 January 1997, 40.

6. DAMS, DEMOCRACY, AND DEVELOPMENT IN TRANSNATIONAL PERSPECTIVE

1. See Jackie Smith et al., eds., *Transnational Social Movements* (1997); Jonathan Fox and David Brown, eds., *The Struggle for Accountability: The World Bank, NGOs, and Grassroots Movements* (Cambridge: MIT Press, 1998); Ann Florini, ed., *The Third Force: The*

Rise of Transnational Civil Society (New York: Carnegie Press, 2000); and Sanjeev Khagram, James V. Riker, and Kathryn Sikkink, eds., *Restructuring World Politics: Transnational Social Movements, Networks and Norms,* (Minneapolis: University of Minnesota Press, 2002).

2. According the Paul Nelson, three overlapping transnational networks organized campaigns against the World Bank during the 1980s: environment and infrastructure (which focused on big dams), poverty, and structural adjustment. See Paul Nelson, "Internationalising Economic and Environmental Policy: Transnational NGO Networks and The World Bank's Expanding Influence," *Millennium: Journal of International Studies* 25, no. 3 (1996): 605–633.

3. Big dams continued to be built in Southern Europe, especially in Spain and Portugal, as well as in Eastern Europe and the former Soviet Union—with corresponding conflicts over these projects—even after construction declined in Western Europe.

4. See Roberto A. Epple, "Popular Control/Democratic Management of River Basins" (paper Presented to the First International Conference of Peoples Affected by Dams, Curitiba, Brazil, 10 March 1997).

5. See A. D. Usher, *Dams as Aid: A Political Economy of Nordic Development Thinking* (London: Routledge, 1997).

6. Interview with Marie Arnould of European Rivers Network, Curitiba, Brazil, 12 March 1997.

7. This account draws from the following sources among others: Robert S. Devine, "The Trouble With Dams," *Atlantic Monthly,* August 1993, 64–71; Tim Palmer, *Endangered Rivers and the Conservation Movement* (Berkeley: University of California Press, 1986); Marc Reisner, *Cadillac Desert: The American West and Its Disappearing Water* (London: Secker and Warbary, 1986); Richard Gottlieb, *Forcing the Spring: The Transformation of the American Environmental Movement* (Washington, D.C.: Earth Island Press, 1993); Anon., "The Big One," *Economist,* 29 March 1997, 27–28; Anon., "Dammed If You Do," *Economist,* 29 March 1997, 28, 33; and interviews conducted by the author.

8. Interview with Patrick McCully, Campaigns Director of International Rivers Network, 15 July 1995.

9. Devine (1993), 65.

10. Daniel P. Beard, Speech to the International Commission on Irrigation and Drainage, Varna, Bulgaria, 18 May 1994, 4.

11. Blair Harden, "U.S. Orders Maine Dam Destroyed: For First Time, Fish Habitat Takes a Priority Over a Hydroelectric Dam," *Washington Post,* 26 November 1997, A1.

12. "Dammed If You Do," 28.

13. Devine, 66.

14. Philip B. Williams, "The Experience of the International Rivers Network 1985–1997" (paper presented to the First International Conference of Peoples Affected by Dams, Curitiba, Brazil, 10 March 1997).

15. Bruce Rich, *Mortgaging the Earth: The World Bank, Environmental Impoverishment, and the Crisis of Development* (Boston: Beacon, 1994), 111–147. See also David Wirth, "Partnership Advocacy in World Bank Environmental Reform," in Fox and Brown (1998).

16. See U.S. Congress, House Committee on Banking, Finance and Urban Affairs, Subcommittee on International Development Institutions and Finance, Session on Environmental Impacts of Multilateral Development Bank-Fund Projects, 98th Congress, 1st Session, 28–29 June 1983.

17. See Nelson (1996), 609–610.

18. Brent Blackwelder, Foreword to *The Social and Environmental Effects of Large Dams,* vol. 1, ed. Edward Goldsmith and Nicholas Hildyard (Cornwall, U.K.: Wadebridge Ecological Centre, 1984).

19. Edward Goldsmith and Nicholas Hildyard, eds., *The Social and Environmental Effects of Large Dams,* vol. 2 (Cornwall, U.K.: Wadebridge Ecological Centre, 1986).

20. Stephen Schwartzman, *Bankrolling Disasters* (Washington, D.C.: Sierra Club, 1986).

21. See "International Dams Newsletter" 1, no. 1 (Winter 1985–86): 1–2. Published by International Rivers Network, Berkeley, California.

22. See "World Rivers Review" 2, no. 6 (November–December 1987), published by International Rivers Network, Berkeley, California.

23. See Owen Lammers, "IRN's Programmatic Evolution" (unpublished manuscript associated with International Rivers Network, 1998). See also International Rivers Network, *San Francisco Declaration* and *Watershed Management Declaration* (Berkeley: IRN, September, 1988).

24. Philip B. Williams, "A Historic Overview of IRN's Mission" (unpublished manuscript in association with International Rivers Network, 1997).

25. Edward Goldsmith and Nicholas Hildyard, eds., *The Social and Environmental Effects of Large Dams,* vol. 3 (Cornwall, U.K.: Wadebridge Ecological Centre, 1992). Although Goldsmith and Hildyard were once again the editors, volume 3 was the product of a collaboration of an international set of reviewers, demonstrating and contributing to the growing transnationalization anti-dam network. Interestingly, there seems to have been no explicit connection between Goldsmith and Hildyard's volumes and the emergence of transnational links in Europe that eventually led to the formation of the European Rivers Network.

26. Wolfgang Pircher, "36,000 Large Dams and Still More Needed" (paper presented at Seventh Biennial Conference of the British Dam Society, University of Stirling, 25 June 1992).

27. Senior managers in several of these transnational corporations confirmed these changes—which were also visible in the changing corporate citizenship, responsibility, and sustainability policies they adopted—in dozens of interviews conducted between 1995–2000. Virtually all of these interviews were granted on the condition of anonymity. For a more general survey of changing corporate policies, see A. Kolk et al., "International Codes of Conduct and Corporate Social Responsibility: Can Transnational Corporations Regulate Themselves?" *Transnational Corporations* 8, no. 1 (1999).

28. While this section focuses on the World Bank, virtually all donors shifted their policies and practices with respect to big dams over the 1980s and 1990s, from the regional development banks such as the Inter-American Development Bank to bilateral donors such as the U.S. Agency for International Development.

29. See Leonard Sklar and Patrick McCully, "Damming the Rivers: The World Bank's Lending for Large Dams," IRN Working Paper Number 5, November 1994.

30. Personal interview with Thayer Scudder (by phone), 29 April 1996. The United Nations did convene a Panel of Experts to prepare a report on integrated river basin development that was published in 1958 and revised in 1970 but did not highlight the severity of problems related to the environment and resettlement. See Department of Economic and Social Affairs, *Integrated River Basin Development: Report of a Panel of Experts* (New York: United Nations, 1958, rev. 1970).

31. Robert Wade, "Greening the Bank: The Struggle over the Environment, 1970–1995," in *The World Bank: Its First Half Century,* ed. Davesh Kapur, John P. Lewis, and Richard Webb (Washington, D.C.: Brookings Institution, 1997), 2:619–620. The analysis of environmental reform at the Bank builds on Wade's essay and numerous interviews conducted by the author.

32. See Robert McNamara, *Address to the U.N. Conference on the Human Environment, Stockholm, Sweden, 8 June 1972* (New York: United Nations, 1972).

33. Wade in Kapur, Lewis, and Webb (1997), 628–629.

34. See Fox and Brown (1998), chap. 1.

35. Quoted in Andrew Gray, "Development Policy—Development Protest: The World Bank, Indigenous Peoples and NGOs," in Fox and Brown, 7.

36. The World Bank, "Social Issues Associated with Involuntary Resettlement in Bank-Financed Projects," in *Operational Manual Statement 2.33* (Washington, D.C.: World Bank, 1980). See Michael Cernea, "Social Science Research and the Crafting of Policy on Population Resettlement," *Knowledge and Power* 6, nos. 3–4 (1993), and "Social Integration and Population Displacement: The Contribution of Social Science," *International Social Science Journal* 143, no. 1 (1995).

37. Gray in Fox and Brown, 9–10.

38. Quoted in Wade (1997), 630.

39. Ibid., 656.

40. Ibid., 634.

41. See Rich (1994), 117–118.

42. Wade, 673.

43. See Office of Environmental and Scientific Affairs, *Tribal Peoples and Economic Development: A Five Year Implementation Review of OMS 2.34 (1982–1986) and Tribal Peoples' Action Plan* (Washington, D.C.: World Bank, 1987), 13.

44. Operation Evaluations Department, "Learning from Narmada," in *OED Précis* (Washington, D.C.: World Bank, May 1995), 7.

45. See Jonathan Fox, "When Does Reform Policy Influence Practice?: Lessons from the Bankwide Resettlement Review," in Fox and Brown (1998). The Bank extended its involuntary resettlement guidelines to all projects in 1990. See World Bank, "Involuntary Resettlement," in *Operational Directive 4.30* (Washington, D.C.: World Bank, 1990).

46. Quoted in Fox of Fox and Brown (1998).

47. Social Policy and Resettlement Division, "Resettlement and Development: The Bankwide Review of Projects Involving Involuntary Resettlement, 1986–1993," in *Environment Department Paper,* no. 032 (Washington, D.C.: World Bank, 1996), 99.

48. Wade, 685–687, 709–710. Environmental staff at the Bank were probably also motivated and empowered by the ostensible importance of the Rio Conference.

49. World Bank, *Bank Procedures, Disclosure of Operational Information, BP 17.50* (Washington, D.C.: World Bank, September 1993), and *The World Bank Inspection Panel, Resolution 93–10* (Washington, D.C.: World Bank, September 1993). For an analysis see Ibrahim F. I. Shiata, *The World Bank Inspection Panel* (Oxford: Oxford University Press, 1994), esp. 9–13, and Lori Udall, "The World Bank and Public Accountability: Has Anything Changed?" in Fox and Brown.

50. See Patrick McCully, *Silenced Rivers: The Ecology and Politics of Large Dams* (London: Zed Books, 1996), 308.

51. Operations Evaluation Department, *The World Bank's Experience With Large Dams—*

A *Preliminary Review of Impacts* (Washington, D.C.: World Bank, 15 August 1996), 3, 12, 15.

52. Ibid., 9–10.

53. Ibid., 3–4.

54. Interviews with Ricardo Montagnier, Executive Director of MAB, Curitiba, Brazil, and Carlos Vanier, Advisor to MAB, Washington, D.C., 12 May 1997.

55. See Aleta Brown, Introduction to *Proceedings: First International Meeting of People Affected by Dams* (Berkeley: International Rivers Network, June 1997), 7.

56. See the various documents presented and generated at the conference found in *Proceedings* just cited in preceding note.

57. See Brown, 7.

58. Interview with Mathato Kthis'ane, Highlands Church Action Group of Lesotho, Curitiba, Brazil, 14 March 1997.

59. Interviews with various participants at The First International Conference of People Affected by Dams, Curitiba, Brazil, 11–14 March 1997.

60. The following is based on "Declaration of Curitiba," in *Proceedings,* 11–13 and personal observation of the declaration drafting process.

61. See IPS Correspondents, "Environment: Protests Mark First International Dam Day," Washington, Interpress Service, 13 March 1998.

62. Interview with Patrick McCully, IRN Campaigns Director, 3 July 1997.

63. See IUCN pamphlet, *World Commission on Dams* (Cambridge: IUCN, 1998).

64. See Stephanie Flanders, "Truce Called in Battle of the Dams," *Financial Times,* 14 April 1997.

65. Patrick McCully, "A Critique of 'The World Bank's Experience with Large Dams': A Preliminary Review of Impacts" (paper presented at the World Bank-IUCN Workshop: Large Dams: Learning from the Past, Looking at the Future, Gland, Switzerland, 10 April 1997).

66. See Flanders (1997).

67. See IUCN, *Large Dams: Learning from the Past, Looking at the Future,* Workshop Proceedings (Cambridge: IUCN, August 1997), and IUCN, *WCD Mandate* (Cambridge: IUCN, February 1998).

68. See the series of e-mail correspondence between the interim secretariat and the reference group of the WCD between November 1997 and February 1998. Interview with Allen Taylor, senior advisor to Minister Kader Asmal, 15 August, 1998.

69. See IUCN (1998).

70. The corruption scandal that engulfed the LHWP in the late 1990s, which was more consequential because of the growing transnational advocacy and global spread of norms on anti-corruption, significantly empowered critics, and the future of the remaining big dams proposed for the larger scheme remained in doubt. Yet, the building of additional big dams remains firmly on the policy agenda in South Africa. On the transnational anti-corruption movement see Frederick Galtung, "A Global Network to Curb Corruption," in Florini (2000).

71. L. David Brown and Jonathan A. Fox, "Accountability within Transnational Coalitions," in Fox and Brown (1998), 476.

72. See Khagram, Riker, and Sikkink, Introduction to Khagram, Riker, and Sikkink (2002).

73. For the skeptical view on transnational movements, see Sidney Tarrow, "Fishnets, In-

ternets, and Catnets: Globalization and Transnational Collective Action," in *Challenging Authority: The Historical Study of Contentious Politics,* ed. Michael Hanagan et al. (Minneapolis: University of Minnesota Press, 1998).

74. See Chetan Kumar, "Transnational Networks and Campaigns for Democracy," in Florini (2000).

75. Export Credit agencies have been extremely important to this "privatization process" of big dam building, since these agencies are now the largest single official source of financing to developing countries. See Malcolm Stephens, *The Changing Role of Export Credit Agencies* (Washington, D.C.: International Monetary Fund, 1999).

76. Ian Pope, "Solving the Project Funding Puzzle," *HRW* (April 1996): 30.

77. Another explanation that proponents of big dams independently changed their approaches to these projects was also investigated during the research and examined at several points in this study. While internal "social learning" among pro-dam actors did certainly occur , it was consistently motivated and shaped by external political pressure and critical advocacy.

78. This is not an unexpected development from the perspective of social movement theory. Contentious politics often generate countermovements by targets or opponents at later phases of cycles of social movements. See Mark Traugott, ed., *Repertoires and Cycles of Collective Action* (Durham, N.C.: Duke University Press, 1995).

79. See for example the publications of ICOLD, IHA, and ICID from the early 1990s onward in which the importance of big dams are much more clearly articulated, for example, Turfan, ed., *Benefits and Concerns about Dams* (Ankara: TRCOLD, 1999). However, interestingly, many of these more recent arguments in favor of big dams defend them on the basis of novel norms such as contributing to environmentally sustainable development. For example, hydropower is argued as a preferred option, because it is an environmentally benign source of energy compared with thermal power plants.

80. ICOLD, "Position Paper on Dams and the Environment" (Paris: ICOLD, 1997).

81. See Kathryn Sikkink, "Conclusions," in Khagram, Riker, and Sikkink; and Brown and Fox (1998). .

82. See Robert Goodland, "The Big Dam Controversy: Killing Hydro Projects Promotes Coal and Nukes" (paper presented at the GTE-Technology and Ethics Series, Michigan Technological University, Houghton, Michigan, 1995).

83. It is crucial to know, for example, whether particular postulated aspects of "good governance" such as human rights records are differently correlated with economic growth as opposed to advances in human capabilities such as longevity or literacy. For emergent conceptualizations and indicators of development, see the various annual Human Development Reports produced by the United Nations Development Program from 1990 on, or the extensive debate over the meaning of sustainable development that was sparked by the World Commission on Environment and Development's adoption of a particular understanding of that notion in its 1986 report.

84. See James Ferguson, *The Anti-Politics Machine: Development, Depoliticization, and Bureaucratic Power in the Third World* (Cambridge: Cambridge University Press, 1990), for a particularly compelling piece of scholarship along these lines.

85. Frederick Cooper and Randall Packard, eds., *International Development and the Social Sciences* (Berkeley: University of California Press, 1997), 2–3.

86. The intense debates during the World Commission on Dams process between 1998 and 2000 over what "development effectiveness" meant and how it could be opera-

tionalized to evaluate past experience with and to help formulate future norms for big dam projects is a stark example of this.

87. See Thomas Sowell, *A Conflict of Visions: Ideological Origins of Political Struggles* (New York: Basic Books, 1987).

88. See James G. March and Johan P. Olsen, "Institutional Dynamics of International Political Orders," in *Exploration and Contestation in the Study of World Politics, ed.* Peter Katzenstein, Robert Keohane, and Stephen Krasner (Cambridge: MIT Press, 1999).

89. 107 See Smith et al. (1997); Fox and Brown (1998); Florini (2000); and Khagram, Riker, and Sikkink (2002).

90. For an example of work along these lines, see John Braithwaitte and Peter Drahos, *Global Business Regulation* (Cambridge: Cambridge University Press, 2000).

91. For a seminal predominantly liberal and actor-centric volume in this recent wave of scholarship, see Thomas Risse-Kappen, ed., *Bringing Transnational Relations Back In* (Cambridge: Cambridge University Press, 1995). This is less so the case for sociological work in the subfield of transnational migration.

92. For one of the premier books in this genre, see John Gerard Ruggie, *Constructing the World Polity: Essays on International Institutionalization* (London: Routledge, 1998).

93. 111 Katzenstein, Keohane, and Krasner (1999), 35.

94. For a definitive article on the world society perspective, see John W. Meyer and John Boli, "World Society and the Nation-State," *American Journal of Sociology* 103, no. 1 (1997). For a recent and compelling book from the world systems approach, see Andre Gunder Frank, *Reorient: Global Economy in the Asian Age* (Berkeley: University of California Berkeley, 1998). For neoinstitutionalist and field theory in sociology, see Walter W. Powell and Paul J. Dimaggio, eds., *The New Institutionalism in Organizational Analysis* (Chicago: University of Chicago Press, 1991), and Pierre Bourdieu, *Outline of a Theory of Practice* (Cambridge: Cambridge University Press, 1977).

95. A virtual landslide of writing on globalization emerged in the 1990s. Much of it focused on the economic aspects of supposed globalization. However, the scholarly literature on globalization ranges from work that is more about internationalization or international integration to that which is about worldwide de-territorialization to that which is about transnational phenomena and dynamics similar to the approach offered in this book. Some of the earlier and more prominent books from a range of theoretical and methodological approaches include Roland Robertson, *Globalisation: Social Theory and Global Culture* (London: Sage, 1992); Arjun Appadurai, *Modernity at Large: Cultural Dimensions of Globalization* (Minneapolis: University of Minnesota Press, 1996); Dani Rodrik, *Has Globalization Gone Too Far?* (Washington, D.C.: Institute for International Economics, 1997); Saskia Sassen, *Globalization and Its Discontents* (New York: New Press, 1998); and David Held et al., *Global Transformations: Politics, Economics, and Culture* (Stanford: Stanford University Press, 1999).

96. For example, research comparing and contrasting the emergence, process, and outcomes of the WCD with other experiments in "global governance" that emerged in the 1990s would be illuminating. Research on novel and emergent transnational institutional arrangements with respect to business-society relations and ethnicity-nation-state dynamics would also be of great theoretical, empirical, and practical value.

INDEX

Note: Page numbers followed by *f* and *t* indicate figures and tables, respectively.

National Resources Defense Council (U.S.), 193
National Wildlife Federation (U.S.), 185, 193
Nayalawla, B. N., 131
Nayanar, E. K., 48
NCEPC. *See* National Committee on Environmental Planning and Coordination (India)
Nehru, Jawaharlal, 35
  and big dam building, 33, 35, 61, 67, 70
  death of, 72
  on displacement of people, 37
Nelson, Paul, 185, 254 n.2
Netherlands Organization for International Development Cooperation (NOVIB), 163
New Molones Dam (U.S.), 181
Nongovernmental organizations (NGOs)
  in anti-dam campaign, 10, 14
  and big dam building, changing dynamics of, 27
  transnational, 11–12, 12t, 13t
  and transnational coalitions and networks, 13
  and transnational political economy of development, 11–12
  and World Bank, 104, 108
  See also specific organizations
Norms and principles, global
  Brazil's adoption of, 147
  compliance pressure on India, 90–96
  and decline in big dam building, 27, 139, 205
  definition of, 222 n.58
  domestic democratization and adoption of, 20
  India's adoption of, 41
  local dynamics and, 224 n.73
  spread and institutionalization of, 3, 15–17
  and transnational anti-dam network, 187
  turning into practices, pathway for, 18–20
  types of, 223 n.61
  and world politics, 222 n.59
Norway, anti-dam campaign in, 179
NOVIB. *See* Netherlands Organization for International Development Cooperation
NWDT. *See* Narmada Water Dispute Tribunal
NWRDC. *See* Narmada Water Resources Development Committee

Official Secrets Act (India), 115–116, 244 n.41
Orange River Basin (South Africa), big dam building in, 165–166
Overseas Development Administration (ODA) (U.K.), 7

Oxfam (U.K.), 59, 91, 188
  and Arch Vahini (India), 91
  and multilateral development banks campaign, 185
  and Narmada Projects, campaign against, 106

Packard, Randall, 212
Paiakan, Paulinho, 156
Paranjipye, Vijay, 116, 118
Partridge, William, 98
Patel, Anil, 90, 91, 92, 93, 95, 98
  letter to World Bank, 89, 240 n.59
Patel, Chimanbhai, 110, 122, 123
Patel, Girish, 79–80, 118
Patel, H. M., 45
Patkar, Medha
  in anti-dam movement, 61, 92
  on development, 1
  fasts by, 121, 123, 132, 134
  first visit to U.S., 109
  meetings with World Bank representatives, 118, 120
  on Narmada Projects
    environmental controversies, 107–108
    resettlement controversies, 114, 115
  and social mobilization, 107, 112, 114
  testimony to U.S. Congress, 118, 119
  on World Commission on Dams, 204
Pawar, Sharad, 122
Pedra do Cavalo Project (Brazil), 144
Peoples Union for Civil Rights (India), 127
Pircher, Wolfgang, 189
Posey, Darrel, 156, 193
Potter, Frank, 234 n.70
Power (energy)
  conflicts over, 7–8
  Indian big dam projects and, 68, 70
Press
  in India, 47, 59
  in Indonesia, 162
  and sustainable development advocacy, 25
Preston, Lewis, 129, 196
Prevention of Air Pollution Act (India, 1972), 42, 103
PRIA. *See* Society for Participatory Research in Asia
Private dam industry, 7
  in India, 63
  transnational anti-dam network and, 208–209
Pro-Indian Commission of São Paulo (CPI-SP), 156, 157

Probe International (Canada), 52, 173, 188, 234
    n.70
Public Interest Research Group (India), 52

Qureshi, Moeen, 117–118, 124

Rajpipla Social Services Society (India), 88
Rao, K. L., 72–73
Rao, Narishima, 58
Reddy, M. S., 120
Regional Commission of People Affected by
    Dams (CRAB) (Brazil), 149, 150–152,
    157
Resettlement and rehabilitation (R&R)
    Narmada Projects and, 76, 80, 82, 88–89,
        111–112, 126, 128, 240 n.59
    World Bank policies on, 88–89, 192, 194
Rich, Bruce, 52, 108, 185, 186, 197
Riker, James V., 160
Rio Conference. *See* United Nations Confer-
    ence on Environment and Develop-
    ment (1992)
Robless, C. L., 94
Roy, Arundhati, 247 n.98
Roy, S. K., 55

Sahniser, Howard, 180
Salam, Abdul, 98
Salim, Emil, 160
Salve, N. K. P., 60–61
San Francisco Declaration (1988), 187
Sardar Sarovar Project (India)
    continued construction of, 131, 132, 136
    displaced peoples from, 82, 87–88
    domestic resistance to, 84, 87–88, 116, 121
    environmental impact study of, 104–105
    height of, 83, 84
    inter-State conflict over, 77–78
    planning for, 86–87
    resettlement reform campaign, 90–99
    stalled construction of, 2, 101
    See also Narmada Projects (India)
Savage, J. L., 67
Save Nimar Action Committee, 84–85
Save Silent Valley Committee, 45, 46, 47
Save the Narmada Movement. *See* Narmada
    Bachao Andolan
Scheuer, James, 118, 119
Schwartzman, Stephen, 143, 186
Scudder, Thayer, 89, 92, 93, 105, 118, 190
Seshan, T. N., 50, 55, 107, 110
Sharma, L. T., 59

Sharma, Ravi, 59
Sharma, Shankar Dayal, 85
Shui Fu, 170
Shukla, V. C., 85, 132
Sierra Club, 180, 183
*Silenced Rivers: The Ecology and Politics of
    Large Dams* (McCully), 199
Silent Valley National Park (India), 48, 49
Silent Valley Project (India), 43–49
    campaign against, 42, 184
    domestic opposition to, 44, 45
    and electoral politics, 46–47
    inter-government conflict over, 47–48
    stalemate over, 48
    transnational criticism of, 43–44, 45, 233 n.51
Singh, Arjun, 85, 118
Singh, Charan, 46
Sitarambhai (Indian tribal leader), 116
Sobradinho Dam (Brazil), 144, 149
    people displaced by, 152–153
*The Social and Environmental Effects of Large
    Dams* (Goldsmith and Hildyard), 59,
    186, 189, 255 n.25
Social mobilization
    degrees of, 29f, 140f
    and development, 20, 21f
    factors responsible for, 20–21
    See also Domestic activism
Society for Participatory Research in Asia
    (PRIA), 59
Soeharto, 159, 160, 162, 164
Soekarno, 159
Solanki, Madhavsingh, 85
SOS Loire Vivante (France), 179
South Africa, 165–170, 206
    big dam building in, 31, 141, 165–166, 169–
        170 (*See also specific projects*)
    democratization of, 168
    political regime in, 29f, 140f, 166, 167
    social mobilization in, 29f, 140f, 169
Soviet Union
    big dams in, 5
    support for Tehri Project (India), 56–57
State(s)
    autonomous vs. porous, 26
    and development, 15
    institutional isomorphism of, 15–16
    norms legitimating, 15, 223 n.62
    resistance to globalizing norms, 19–20
    use of term, 228 n.122
    weak vs. strong, 26
Stegner, Wallace, 180